MW00358464

ROLEX®
HIGHLIGHTS

HERBERT JAMES

4880 Lower Valley Road • Atglen, PA 19310

Other Schiffer Books by the author:

Omega Highlights. ISBN: 978-0-7643-4212-7. $29.99
A. Lange & Söhne Highlights.
ISBN: 978-0-7643-4361-2. $29.99
Breitling Highlights. ISBN: 978-0-7643-4211-0. $29.99

Other Schiffer Books on Related Subjects:

Breitling: The History of a Great Brand of Watches 1884 to the Present. Benno Richter. ISBN: 9780764326707. $49.95
Omega Designs: Feast for the Eyes. Anton Kreuzer.
ISBN: 9780764329951. $59.99
Rolex Wristwatches: An Unauthorized History. James M. Dowling & Jeffrey P. Hess. ISBN: 0764324373. $125.00
Swiss Wristwatches: Chronology of Worldwide Success. Gisbert L. Brunner & Christian Pfeiffer-Belli.
ISBN: 0887403018. $69.95
Vintage Rolex® Sports Models - 3rd Edition.
Martin Skeet & Nick Urul. ISBN: 9780764329814. $79.99
Wristwatch Chronometers. Fritz von Osterhausen.
ISBN: 0764303759. $79.95

Originally published as *Rolex Highlights* by HEEL Verlag GmbH, Königswinter © 2013: HEEL Verlag GmbH
Picture Credits:
Archives of the publisher, Auktionshaus Dr. Crott, Rolex

Layout and Design:
Muser Medien GmbH, Mannheim
Christine Mertens

Lithography:
Muser Medien GmbH, Mannheim
Fred Klöpfel

Translated from the German
by Omicron Language Solutions, LLC

Library of Congress Control Number: 2014948367

Cover by Justin Watkinson
Type set in Tall Films Expanded /Avant Garde Gothic Itc

ISBN: 978-0-7643-4684-2
Printed in China

Published by Schiffer Publishing, Ltd.
4880 Lower Valley Road
Atglen, PA 19310
Phone: (610) 593-1777; Fax: (610) 593-2002
E-mail: Info@schifferbooks.com

ROLEX

CONTENTS

ROLEX

FOREWORD

Rolex—no other watch brand has a mightier reputation. The name has long been synonymous with luxurious wristwatches of Swiss origin. Those who wear a Rolex have made it in life. Those who wear a Rolex are wealthy and like to show it. Everyone knows this. And the Rolex legend is more alive than ever in the twenty-first century.

Often imitated, the Rolex is rarely equaled. In fact, no other brand has contributed as much to the development of the wristwatch than the manufacturer from Geneva. From the pioneering achievements of the first Oyster models and the design of the automatic movement to the extraordinary high-performance chronometers of its Professional line, made since the 1950s—all of these innovations have made Rolex the trailblazer of an entire industry. Over decades, the collection of these top watches was optimized and improved. Rolex has created its own world, independent from the undulations of the watch industry, a world of fascination that attracts hundred of thousands of customers every year.

This book will give you an overview of what Rolex has built since its founding by Hans Wilsdorf in 1908. We cast a glance at the history of the brand, which has been almost constantly characterized by success and expansion, and we introduce you to the most important models from its history, primarily from the current collection, with beautifully detailed pictures and technical specifications. Within this context, it goes without saying that not all of the numerous versions and references can be presented here. But maybe reading *Rolex Highlights* will arouse your interest and you will dig deeper into specialist publications about the Rolex brand, or to find your local jeweler and become a part of the legend yourself.

Herbert James

THE ROLEX STORY— A CHRONOLOGY OF SUCCESS

The story of the most famous Swiss watch brand begins in Kulmbach, Germany. Hans Wilsdorf, the man whose work and wealth of ideas would play a decisive role in the vaguely foreseeable history of wristwatches, was born here in 1881. Wilsdorf lost both of his parents when he was young and he initially completed a vocational training for a commercial occupation. At the turn of the century he left Bavaria to settle in the Canton of Jura in Switzerland. It wasn't just the scenery that brought him here, but La Chaux-de-Fonds, the Mecca of Swiss watchmaking, home to numerous brands, steeped in tradition, and the work site of the best watchmakers in the world. When he first arrived, he worked in the exporting of watches. In 1903, though, he moved to London—the center of the Western world at the time—to import Swiss pocket watches that had had their timing accuracy certified by the observatory in Neuchâtel.

Hans Wilsdorf

From Wilsdorf & Davis to Rolex

Two years later, he decided to go into business for himself and founded the watch wholesaler Wilsdorf & Davis with case builder Alfred Davis. After he had observed the traditional watch market for a few years (the pocket watch was well established and common in the beginning of the 20th century; wristwatches were only worn as expensive jewelry by wealthy women and were ridiculed by the men of the watch industry), he came to the conclusion that the future would belong to the wristwatch. As a useful everyday object for both men and women, they were superior to the impractical pocket watch, but they were still not particularly accurate.

The first Rolex with chronometer certificate from 1910.

Early on, Wilsdorf had recognized precision, water-resistance, robustness, and versatility in all situations as the core qualities for the success of the wristwatch, and their establishment paved the way for his own watches. Thus, he had already anticipated everything that Rolex would later become—a foresight that nobody else in the watch industry possessed. In 1908, the name Rolex (short for "Rolling Export") was entered in the brand name registry.

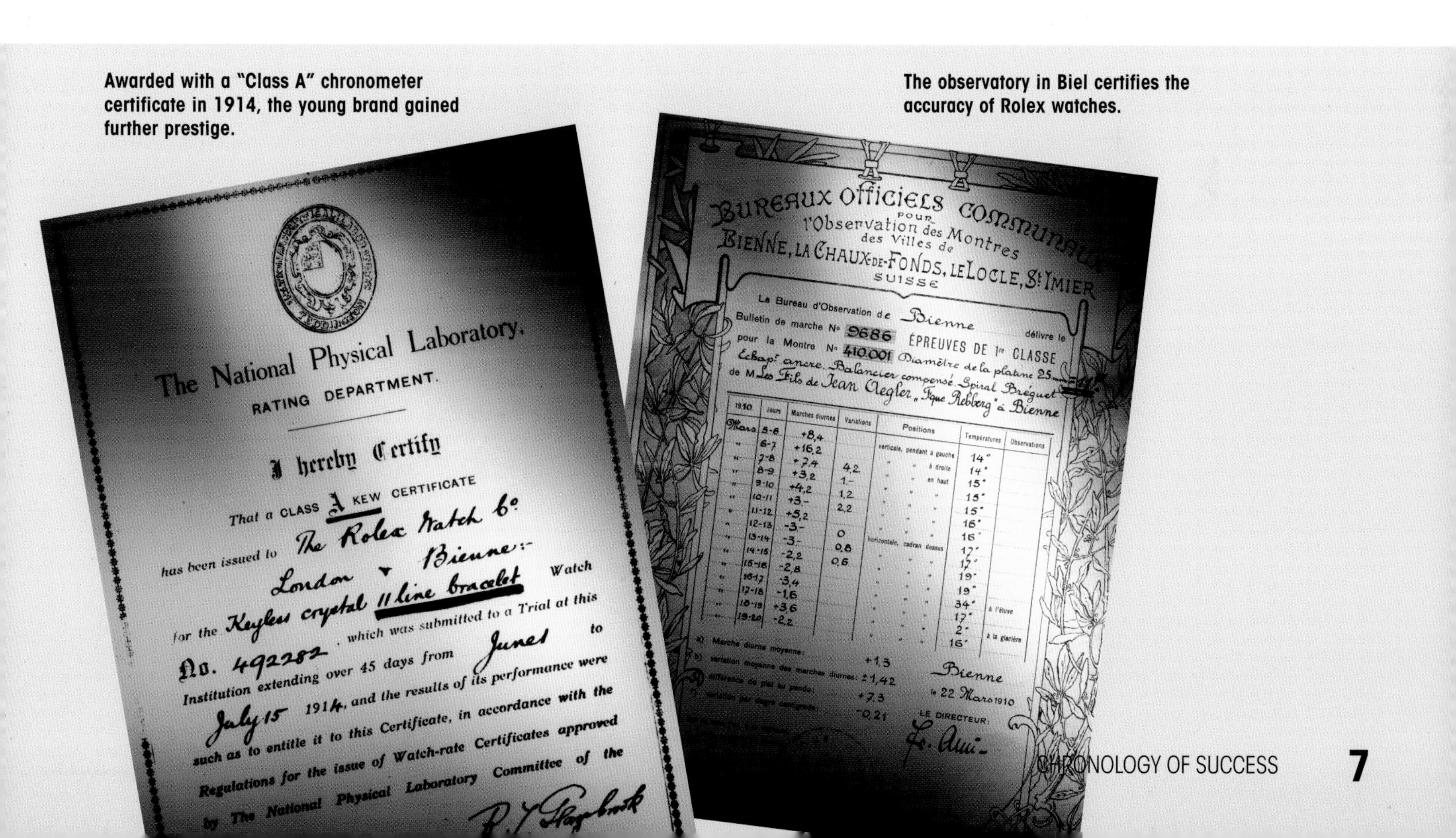

The National Physical Laboratory,
RATING DEPARTMENT.

I hereby Certify

That a CLASS A KEW CERTIFICATE has been issued to The Rolex Watch Co London & Bienne:- for the Keyless crystal 11 line bracelet Watch No. 492282, which was submitted to a Trial at this Institution extending over 45 days from June 1 to July 15 1914, and the results of its performance were such as to entitle it to this Certificate, in accordance with the Regulations for the issue of Watch-rate Certificates approved by The National Physical Laboratory Committee of the

Bureaux Officiels Communaux pour l'Observation des Montres des Villes de Bienne, La Chaux-de-Fonds, Le Locle, St Imier
Suisse

Le Bureau d'Observation de Bienne délivre le Bulletin de marche No 9686 ÉPREUVES DE 1re CLASSE pour la Montre No 410.001 Diamètre de la platine 25 mm Echapt ancre. Balancier compensé. Spiral Bréguet de M Les Fils de Jean Aegler „Fque Rebberg" à Bienne

1910	Jours	Marches diurnes	Variations	Positions	Températures	Observations
Mars	5-6	+8,4		verticale, pendant à gauche	14°	
"	6-7	+16,2		" " à droite	14°	
"	7-8	+7,4		" " en haut	15°	
"	8-9	+3,2	4,2	" " "	15°	
"	9-10	+4,2	1.-	" " "	15°	
"	10-11	+3.-	1,2	" " "	16°	
"	11-12	+5,2	2,2	" " "	16°	
"	12-13	-3.-		horizontale, cadran dessus	17°	
"	13-14	-3.-	0	" " "	17°	
"	14-15	-2,2	0,8	" " "	19°	
"	15-16	-2,8	0,6	" " "	19°	
"	16-17	-3,4		" " "	34°	à l'étuve
"	17-18	-1,6		" " "	17°	
"	18-19	+3,6		" " "	2°	à la glacière
"	19-20	-2,2		" " "	16°	

a) Marche diurne moyenne: +1,3
b) variation moyenne des marches diurnes: ±1,42
c) différence du plat au pendu: +7,3
d) variation par degré centigrade: -0,21

Bienne le 22 Mars 1910

LE DIRECTEUR:

Awarded with a "Class A" chronometer certificate in 1914, the young brand gained further prestige.

The observatory in Biel certifies the accuracy of Rolex watches.

FOR KING AND COUNTRY

Rolex introduces for the
the greatest Triumph in Watch

ROLEX 'OYSTE

The Wonder Watch that Defies the Elements.

MOISTURE PROOF
WATER PROOF
HEAT PROOF
VIBRATION PROOF
COLD PROOF
DUST PROOF

BEING hermetically sealed the Rolex 'Oyster' is proof against changes of climate, dust, water, damp, heat, moisture, cold, sand or grease; it can, in consequence, be worn in the sea or bath without injury, nor would arctic or tropical conditions affect the wonderful precision of its beautifully poised movement. The introduction of the Rolex 'Oyster' model marks an unique development in the forward stride of the chronometric science, and perfect timekeeping under all conditions is at last a possibility.

Copyright by Pacific and Atlantic Photos, Ltd.

Miss Mercedes Gleitze carried an 'Oyster' throughout her recent Channel Swim. More than ten hours of submersion under the most trying conditions failed to harm its perfect timekeeping. No moisture had penetrated and not the slightest corrosion or condensation was revealed in the subsequent examination of the Watch.

A HANDSOMELY printed, fully informative Brochure illustrating the entire range of OYSTER Models available, together with the name of nearest jeweller stocking ROLEX Watches, sent post free to any reader who makes application by post-card or letter, giving name and address to our London Office:

THE ROLEX WATCH CO. Ltd.
40/44,
Holborn Viaduct,
London, E.C.4.

Send for this coloured Brochure it's FREE

ROLEX
OYSTER

THE ROLEX OYSTER

For MEN or WOMEN.

THE ROLEX WATCH CO. LTD.

Mercedes Gleitze and the Oyster on the front page of the Daily Mail in 1927.

The first Oyster from 1926.

Short, simple to pronounce, and easy to remember, the name derived from Wilsdorf's conclusion that a strong presence and high recognition value of his unique watches would be firmly fixed in people's minds. The Franconian by birth had now gone from pure dealer to owner of his own watch brand, with international distribution and marketing, at twenty-seven years of age!

The components of his watches were supplied by Swiss manufacturers, among them Maison Aegler in Biel. Wilsdorf believed that this watchmaker workshop was the only one that was capable of producing the small, indispensable movements for wristwatches at that time. He needed them for the British market and, thus, for the entire British Empire. Ultimately, he wanted to prove to all of his skeptical contemporaries that a wristwatch could have the accuracy of a chronometer.

By 1910, he had attained his goal: the first Rolex wristwatch had received a chronometer certificate. Four years later, a similar model obtained a "Class A" certificate from the Kew Observatory in England; this was a seal of approval of the highest order, since these certificates had previously been reserved solely for marine chronometers. The accuracy of Wilsdorf's wristwatches had been proven. Next, he would work intensely to make his watches water-resistant, which would in turn make them more practical for everyday use.

In 1914, after the outbreak of the First World War had rapidly increased import duties, he moved company headquarters to Biel. In 1919, after the war and all of the upheaval in Europe, he left England entirely and settled down in Geneva, where he founded the company Montres Rolex SA one year later.

Right: The first automatic movement from 1931.

ROLEX PERPETUAL
CHRONOMETER
N 9401
SWISS MADE
SUPER - PATENTED BALANCE
VIS TIRETTE
TIMED 6 POSITIONS
ROLEX
CHRONOMETER
18 RUBIES
SWISS MADE

The first Datejust from 1945.

Now he was closer to the manufacturer of his watch components and operating in the heart of the Swiss watch industry. He could not have chosen a better location: he profited from the centuries-old tradition of fine craftsmanship and the international reputation of the cosmopolitan city of Geneva.

The Rolex Oyster—A Breakthrough

Wilsdof's quest for a water-resistant case found success in 1926, when he presented the world's first waterproof watch: the Rolex Oyster. The case had a screw-down bezel and case back, and the crown could also be screwed into the side of the case. This hermetically sealed case offered optimal protection for the movement—Rolex had again managed to get a head start on the fast-approaching competition of the quickly expanding wristwatch industry. Meanwhile, word was spreading and people were catching on to the fact that a timepiece worn on the wrist was more practical and more suitable for daily use than a pocket watch.

But Wilsdorf didn't rest on his laurels. He knew that he had to defend this competitive edge and wanted Rolex to continue to be perceived as an innovator and a pioneer of an entire industry.

Sir Edmund Hillary wore an Explorer during the first ascent of Mount Everest.

The Rolex Explorer was launched in 1953.

Left: Watch production at Rolex was operating at full speed in the postwar years.

The Submariner is considered the mother of all diver's watches.

The GMT-Master debuted in 1955.

The companies Aegler and Rolex in Biel.

His flair for effective marketing prompted him to support the young Englishwoman Mercedes Gleitze with her plans to swim across the English Channel in 1927. He outfitted the long-distance swimmer with an Oyster and, when she emerged from the tides of the channel more than ten hours later, it was still functioning. It had withstood the stresses of swimming and the exposure to the salt water. Rolex announced this success in a full-page advertisement on the front page of the *Daily Mail*, claiming the "greatest triumph in watchmaking." It was also the first testimonial, the first time that a prominent person had promoted a watch brand—with resounding success: Rolex and its Oyster achieved enormous recognition and quickly became synonymous with dependable, water-resistant wristwatches of Swiss provenance.

Hans Wilsdorf passed away in 1960 at the age of 79.

Rolex dove down into the Mariana Trench aboard the Trieste.

The first Cosmograph Daytona from 1960.

Automatic Winding and Testimonials

The next milestone in the company's history was not far away. Wilsdorf soon went to work developing the Oyster even further. Winding the watch by hand every day would loosen the crown, which would in turn threaten the water-resistance of the watch. To prevent this, Wilsdorf had a rotor mechanism developed that would swing in both directions; the mainspring of the movement remained wound as a result of the natural motion of the wearer's arm. Winding of the movement over the crown would become completely unnecessary, provided that the watch was worn every day. In the 1930s, more features were added that are mainstays of Rolex watches today. The crown was officially registered as the brand's logo in 1931 (in the thirties, it was found on the dial; beginning in the fifties, on the crown). Rolesor, an alloy of gold and stainless steel, and another distinguishing feature of Rolex watches, has been a part of Rolex nomenclature since 1933.

Rolex headquarters in Geneva.

After World War II, Wilsdorf finally began to cash in on his technical innovations and his knack for communication, marketing, and emerging trends. With innovations like the Datejust in 1945 and the DayDate in 1956, Wilsdorf paved the way to the future. Meanwhile, more and more celebrities were wearing Rolex and the brand's prestige increased relentlessly. Thanks to the company's cleverly cultivated liaison with race car drivers, pilots, and explorers, Rolex watches established themselves as guarantors for reliably performing their duties, even in absolutely extreme situations. Whether it was Sir Malcolm Campbell, who set a land speed record of 300 mph (480 km/h), Chuck Yeager, who was the first person to break the sound barrier in an airplane (1947), or even Sir Edmund Hillary during the first ascent of Mount Everest in 1953—they all wore an Oyster. And Hans Wilsdorf, of course, didn't fail to inform the entire world about it.

The Classics of the Postwar Years

Little by little, watches that were specifically constructed for extreme situations were accepted into the slowly expanding Rolex collection. During the 1950s, in particular, pioneering models emerged that continue to be perfected today. Decades later their aura remains intact.

The Explorer, introduced in 1953, was on Mount Everest. The 1955 Submariner was the first real diver's watch and was waterproof up to 200 meters (20 bar pressure). The GMT-Master, made primarily for pilots who moved through time zones frequently and wanted to read different times simultaneously, was also launched in 1955.

One year later the Milgauss was launched, which was designed to resist magnetic fields, and, in 1963, the chronograph Cosmograph Daytona was born, which could measure average speeds with its tachymeter bezel. In 1960, the extreme sport watch Deep Sea Special made it to the deepest part of the ocean (10,916 meters) in the Mariana Trench, attached to the outside of the deep-sea submersible *Trieste* captained by Jacques Picard. The Sea-Dweller (1967) also stood in this tradition and was made for diving in extreme depths.
Hans Wilsdorf passed away in 1960 after successfully leading his company for over fifty years. The company and its assets were transferred to the Hans Wilsdorf Foundation and Chairman André Heiniger continued the work of the founder in 1963. Building upon the solid foundation, his clever model policy and marketing activities made Rolex the most popular watch brand in the world. He was equal to his predecessor in his foresight and his sense for market developments.

The Quartz Crisis and Global Success

When quartz technology made sure that mechanical watches would be seen as relics of days gone by in the 1970s, Rolex took part in developing the first Swiss-produced quartz movement. A Rolex with this new, battery-powered movement actually appeared in 1977: the Oysterquartz.

ROLEX

Rolex decided not to continue along this path, though, concentrating instead on perfecting the high-performance automatic watches that had shaped the brand for decades. With models like the Explorer II, the Sea Dweller 4000, and the GMT-Master II, the existing collection was developed and expanded.

At the same time, Rolex strengthened its efforts in sponsorship. Researchers and sportsmen, in particular, took center stage in promoting the Geneva company. Golf, tennis, motor sports, and professional sailing events were increasingly being held under the crown logo of the Swiss brand. Despite the triumphant march of quartz watches, Rolex's prestige remained virtually untouched. This success is, of course, owed to the products whose quality continued to grow, but primarily to the campaigns that Rolex continues to lead with vision and skill.

The Yacht-Master made its debut in 1992 and the women's collection was expanded with its Pearlmaster models. Patrick Heiniger followed in his father's footsteps by resuming Rolex's "vertical integration" plan: Rolex wanted to make itself completely independent of other manufacturers and produce everything on its own, from the smallest movement parts to its cases. The plan has proven to be a success and, today, Rolex is one of the few manufacturers that can produce its own movements, with their many components, completely independent of other companies. One result to admire was the new chronograph movement 4130, which Rolex installed in the Cosmograph Daytona beginning in 2000.

Rolex was better positioned for the ensuing watch boom of the 21st century than many other watch brands and used this circumstance for unprecedented expansion and development of its model families, such as the DeepSea, Explorer II, Milgauss, and the new Sky-Dweller. And there is no end in sight for Rolex's success.

In the following chapters, you'll find a selection of the best watches from the current collection and a description of their most outstanding features.

Left: Today Rolex manufactures all of its own movement components.

ROLEX TODAY

Rolex remains true to the ideas of founder Hans Wilsdorf. Well into the 21st century, design, aesthetics, functionality, and marketing are still the standards which the visionary had established for his brand in the decades before and after World War II, thanks primarily to the Hans Wilsdorf Foundation, which has owned 100% of the company since his death. Gian Marco Marini has been in charge of Rolex since 2011, just the fourth president in the company's 115 year history.

Independent and Forward-Looking

While many competitors manufacture their components in East Asia in the 21st century, Rolex production takes place exclusively at four facilities in Switzerland, with more than 6,000 active employees. Over a century after the brand was founded, the company headquarters is still located in Geneva. Administration of the company takes place there too, as do final assembly, quality control, sales, and customer service. In Chêne-Bourg, dials and jeweled watch components are designed and produced, while cases and bracelets are designed, produced, and inspected in Plan-les-Ouates, in black, glass buildings newly constructed in 2008.
Biel, the former headquarters of the brand from 1914 to 1919, still has a prominent position within the company structure.

All Rolex movements are manufactured and assembled in a facility that opened there after three years of construction in October 2012 to complement other production facilities. Around 2,000 employees are busy designing, making, and assembling movement blanks, balance springs, and escapement components.

This integrated production infrastructure allows the brand to be almost completely independent of third-party manufacturers. From the smelting of various gold alloys to the assembly of movement components—everything is made in-house according to strict guidelines. More importantly, though, it makes chronometer precision possible, which is more difficult to attain in mass production than in time-consuming single-unit production.

Rolex still produces by far the most chronometer-certified mechanical watches. In 2011, around 751,000 of their movements satisfied the stringent criteria of the testing laboratory COSC (Contrôle officiel suisse des chronomètres).

Global Player and Supporter

In 2013, Rolex is represented by distributors in around 100 countries worldwide. Thirty branch offices and customer service centers guarantee that this network of specialist dealers offer quality and authenticity for all Rolex wristwatches. This is more important than ever now, since being the most prestigious watch brand in the world means a myriad of cheap imitations made in the Far East.

ROLEX

Approximately 4,000 Rolex-trained watchmakers work in branch offices and at specialist dealers, where they handle maintenance and repairs. Keeping with the tradition of founder Hans Wilsdorf, Rolex operates an extensive sponsorship program in sports, art, and culture. The brand sponsors one of the best tennis players of all time in Roger Federer. In the world of alpine skiing, and in the tradition of Frenchman Jean-Claude Killy, Rolex is very excited about brand ambassador Lindsay Vonn. The brand is also represented at big international equestrian tournaments.

Rolex has been involved with the sport of golf for more than fifty years and currently sponsors top golfers Tiger Woods and Martin Kaymer. You will also encounter the brand with the crown emblem in sailing and motor sports. The Swiss manufacturer is also at home in the world of opera: Placido Domingo and Cecilia Bartoli, two of the most famous voices worldwide, are signed on, while Canadian Michael Bublé, a contemporary representative of jazz, also gives his endorsement.

In 2013, Rolex cemented its strong position in the watch industry with its impressive showing at Baselworld, the world's biggest trade fair for watches and jewelry. It has grown from a small trade fair booth on just 30 square meters at its first appearance in Basel in 1939 to 1,230 square meters of floor space spread over three stories for exhibiting new models and conducting information sessions and sales talks—

a symbol of global significance and outstanding status for the watch manufacturer from Geneva.

At a time when the image of a brand is more important than ever before, Rolex has achieved unparalleled status in the world of watches. The Rolex legend is more alive than ever, and there is every indication that it will remain so in the decades to come—defying all contemporary trends and changes.

ROLEX

HISTORICALLY SIGNIFICANT MODELS

No other brand is as highly sought-after by collectors as Rolex. This is partly because of its present-ay mystique, but mainly is due to the fact that throughout its history it has produced countless limited and very limited series of exclusive timekeepers that fetch maximum prices at auctions and collectors' fairs due to their rarity value and their technical innovations in mechanics and case construction. Those who buy a Rolex today hope that it will increase in value, or at least maintain its value for many years. Those who already own a classic Rolex wristwatch, on the contrary, can be sure that they will get considerably more for it than they once invested in the timepiece.

CHRONOGRAPH ANTI-MAGNÉTIQUE (1926)

Reference: 2057
Movement: Hand-wound, column wheel control, finely ground and beveled chronograph steel parts and polished screws
Functions: Hours, minutes, auxiliary seconds; chronograph
Case: Yellow gold, 32 x 37 mm; three-body, push back
Remarks: Very rare crown pusher chronograph with 30-minute counter
Estimated Value: $32,000 (€ 24,000)

PRINCE "1/4 CENTURY CLUB" (1930)

Reference: 3937
Movement: Hand-wound, rhodium-plated
Functions: Hours, minutes, auxiliary seconds
Case: Yellow gold, 22 x 44 mm; push back
Estimated Value: $6700 (€ 5000)

TECHNICAL DATA

PRINCE RAILWAY JUMPING HOURS "OBSERVATORY QUALITY" (1930)

Reference: 1587HS
Movement: Hand-wound, rhodium-plated
Functions: Hours (digital, jumping), minutes, auxiliary seconds
Case: Rose and white gold, 23 x 42 mm; push back
Remarks: Rare men's watch with tiered rectangular case
Estimated Value: $24,000 (€ 18,000)

MILITARY ANTIMAGNETIC CHRONOGRAPH (1932)

Reference: 2508
Movement: Hand-wound, column wheel control, finely ground chronograph steel parts and polished screws; nickel-plated
Functions: Hours, minutes, auxiliary seconds; Chronograph
Case: Stainless steel, Ø 35 mm; three-body, push back
Remarks: Rare chronograph with 30-minute counter, one of the first chronographs produced by Rolex
Estimated Value: $27,000 (€ 20,000)

OYSTER CHRONOMETER (1932)

Reference: 3474
Movement: Hand-wound, rhodium-plated
Functions: Hours, minutes, auxiliary seconds
Case: Yellow gold, 32 x 35 mm; screw back
Remarks: Early, water-resistant Oyster
Estimated Value: $3400 (€ 2500)

CHRONOGRAPH ANTI-MAGNETIC (1934)

Reference: 2508
Movement: Hand-wound, column wheel control; rhodium-plated
Functions: Hours, minutes, auxiliary seconds; chronograph
Case: Yellow gold, Ø 36 mm; push back
Remarks: One of the first Rolex chronographs with a 30-minute counter and tachymeter scale
Estimated Value: $18,800 (€ 14,000)

TECHNICAL DATA

OYSTER "CHANNEL SWIMMER" (1935)

Reference: 3224
Movement: Hand-wound, rhodium-plated
Functions: Hours, minutes, auxiliary seconds
Case: Silver, Ø 33 mm; screw back
Remarks: This watch is reminiscent of the one that Mercedes Gleitze wore when she swam the English Channel in 1927
Estimated Value: $2700 (€ 2000)

CHRONOGRAPH ANTI-MAGNETIC (1945)

Reference: 3525
Movement: Hand-wound, column wheel control; rhodium-plated
Functions: Hours, minutes, auxiliary seconds; chronograph
Case: Yellow gold, Ø 35 mm; screw back
Remarks: One of the first Oyster chronographs with 30-minute counter and tachymeter scale
Estimated Value: $33,600 (€ 25,000)

OYSTER CHRONOGRAPH (1947)

Reference: 4500
Movement: Hand-wound, caliber R23; column wheel control; rhodium-plated
Functions: Hours, minutes, auxiliary seconds; Chronograph
Case: Stainless steel, Ø 36 mm; screw back
Remarks: Rare Oyster chronograph with 30-minute counter and tachymeter scale
Estimated Value: $29,500 (€ 22,000)

OYSTER PERPETUAL CHRONOMETER "MICKEY MOUSE" (1949)

Reference: 3131
Movement: Automatic, rhodium-plated
Functions: Hours, minutes, center seconds
Case: Rose gold, Ø 32 mm; screw back
Remarks: Rare Oyster Perpetual with "Mickey Mouse" dial
Estimated Value: $8000 (€ 6000)

TECHNICAL DATA

OYSTER PERPETUAL DATEJUST CHRONOMETER "BIG BUBBLE BACK" (1951)

Reference: 6075
Movement: Automatic, rhodium-plated
Functions: Hours, minutes, center seconds; date
Case: Yellow gold, Ø 35 mm; screw back
Estimated Value: $18,800 (€ 14,000)

OYSTER CHRONOGRAPH "JEAN-CLAUDE KILLY" (1953)

Reference: 6236
Movement: Hand-wound, caliber 72C, rhodium-plated, with column wheel control, finely ground, beveled chronograph steel parts
Functions: Hours, minutes, auxiliary seconds; chronograph; full calendar with date, weekday and month
Case: Yellow gold, Ø 36 mm; three-body, screw back
Remarks: Extremely rare chronograph with full calendar; only 170 copies were produced
Estimated Value: $107,000 (€ 80,000)

OYSTER PERPETUAL CHRONOMETER "OFFICIALLY CERTIFIED" (1953)

Reference:	6062
Movement:	Automatic, rhodium-plated
Functions:	Hours, minutes, auxiliary seconds; full calendar with date, weekday, month, moon phases
Case:	Stainless steel, Ø 35 mm; screw back
Remarks:	Extremely rare, only 350 copies in yellow gold were produced between 1950 and 1953, 50 in rose gold and a scattered few in stainless steel
Estimated Value:	$121,000 (€ 90,000)

OYSTER PERPETUAL CHRONOMETER PRECISION (1950)

Reference:	8171
Movement:	Automatic, rhodium-plated
Functions:	Hours, minutes, auxiliary seconds; full calendar with date, weekday, month, moon phases
Case:	Stainless steel, Ø 38 mm; screw back
Remarks:	This model was produced in a limited series and is extremely rare
Estimated Value:	$94,000 (€ 70,000)

TECHNICAL DATA

OYSTER PERPETUAL "TRU-BEAT" SUPERLATIVE CHRONOMETER "BY OFFICIAL TEST" (1956)

Reference: 6556
Movement: Automatic, caliber 1040; rhodium-plated
Functions: Hours, minutes, center seconds (jumping)
Case: Stainless steel, Ø 35 mm; screw back
Remarks: Extremely rare Oyster with dial imprint "by official test"
Estimated Value: $9400 (€ 7000)

OYSTER PERPETUAL "MILGAUSS" SUPERLATIVE CHRONOMETER "OFFICIALLY CERTIFIED" (1958)

Reference: 6541
Movement: Automatic, caliber 1066M; rhodium-plated
Functions: Hours, minutes, center seconds
Case: Stainless steel, Ø 36 mm; bezel unidirectional rotation with 60-minute graduation; screw back
Remarks: Extremely rare first generation Milgauss
Estimated Value: $40,300 (€ 30,000)

OYSTER PERPETUAL SUBMARINER 200M "JAMES BOND" (1958)

Reference: 6538
Movement: Automatic, caliber 1030; rhodium-plated
Functions: Hours, minutes, center seconds
Case: Stainless steel, Ø 37 mm; bezel unidirectional rotation with 60-minute graduation; screw back
Remarks: The Submariner owes its byname to its appearances in various early James Bond films
Estimated Value: $33,600 (€ 25,000)

OYSTER COSMOGRAPH DAYTONA "PAUL NEWMAN" (1960)

Reference: 6240
Movement: Hand-wound, column wheel control; rhodium-plated
Functions: Hours, minutes, auxiliary seconds; chronograph
Case: Stainless steel, Ø 37 mm; bezel with tachymeter scale, screw back
Estimated Value: $74,000 (€ 55,000)

CHRONOGRAPH ("PRE-DAYTONA") (1965)

Reference: 6238
Movement: Hand-wound, caliber 72B; column wheel control; rhodium-plated
Functions: Hours, minutes, auxiliary seconds; Chronograph
Case: Yellow gold, Ø 32 mm; screw back
Estimated Value: $67,000 (€ 50,000)

CHRONOGRAPH ("PRE-DAYTONA") (1966)

Reference: 6238
Movement: Hand-wound, caliber 722.1; column wheel control; rhodium-plated
Functions: Hours, minutes, auxiliary seconds; chronograph
Case: Stainless steel, Ø 36 mm; screw back
Estimated Value: $16,800 (€ 20,000)

OYSTER PERPETUAL GMT-MASTER (1966)

Reference: 1675
Movement: Automatic, rhodium-plated
Functions: Hours, minutes, center seconds; date; 24-hour hand (second time zone)
Case: Yellow gold, Ø 39 mm; bezel unidirectional rotation with 24-hour graduation; screw back
Estimated Value: $40,300 (€ 30,000)

OYSTER COSMOGRAPH DAYTONA (1968)

Reference: 6239/6263
Movement: Hand-wound, caliber 727; column wheel control, rhodium-plated
Functions: Hours, minutes, auxiliary seconds; chronograph
Case: Stainless steel, Ø 37 mm; bezel with tachymeter scale, screw back
Estimated Value: $23,000 (€ 17,000)

TECHNICAL DATA

OYSTER PERPETUAL SUBMARINER (1970)

Reference: 5513
Movement: Automatic, rhodium-plated, ground, polished screws
Functions: Hours, minutes, center seconds
Case: Stainless steel, Ø 39 mm; bezel unidirectional rotation with 60-minute graduation; screw back
Remarks: Extremely rare military diving watch of the British Special Boat Service of the Royal Marines; distinguishing features are the "sword" hands
Estimated Value: $40,300 (€ 30,000)

OYSTER PERPETUAL DATE EXPLORER II (1970)

Reference: 1655
Movement: Automatic, caliber 1570; rhodium-plated, polished screws
Functions: Hours, minutes, center seconds; date; 24-hour hand (second time zone)
Case: Stainless steel, Ø 39 mm; bezel unidirectional rotation with 24-hour graduation; screw back
Remarks: rare Explorer II, so-called "Steve McQueen", with Twinlock crown and prominent hand for displaying second time zone
Estimated Value: $13,400 (€ 10,000)

OYSTER SUPERLATIVE CHRONOMETER COSMOGRAPH DAYTONA (1977)

Reference: 6263
Movement: Hand-wound, caliber 727/1531; column wheel control; ground, beveled steel parts; rhodium-plated
Functions: Hours, minutes, auxiliary seconds; Chronograph
Case: Yellow gold, Ø 38 mm; three-body; bezel with tachymeter scale, screw back, screw-down crown and pushers
Estimated Value: $47,000 € 35,000)

OYSTER PERPETUAL DATE EXPLORER II (1984)

Reference: 16550
Movement: Automatic, caliber 3085; rhodium-plated, ground, polished screws
Functions: Hours, minutes, center seconds; date; 24-hour hand (second time zone)
Case: Stainless steel, Ø 39 mm; three-body; bezel unidirectional rotation with 24-hour graduation; screw back
Remarks: Explorer II with rare cream-colored dial
Estimated Value: $8000 (€ 6000)

ROLEX GMT-MASTER II

It's fair to say that the GMT-Master II is the founding father of all time zone watches. In the last few decades, the number of people who frequently travel to different time zones has increased dramatically. Rolex realized in the mid-1950s that it's extremely practical to be able to keep an eye on your home time when taking short trips around the world—primarily, of course, for pilots of airlines that operate worldwide. The GMT-Master initially had a 24-hour hand connected to the hour hand. Since its redesign in 1985, though, it can now be adjusted to the respective time zone completely independent of the main time. In its last major overhaul in 2006, the movement was perfected even further and made more robust so that the GMT-Master reliably performs its duty anywhere, anytime and in any time zone.

TECHNICAL DATA

OYSTER PERPETUAL GMT-MASTER II

Reference:	116710BLNR
Movement:	Automatic, Rolex caliber 3186; Ø 28.5 mm, height 6.4 mm; 31 jewels; 28,800 A/h; certified chronometer (COSC)
Functions:	Hours, minutes, center seconds; additional 24-hour display; date
Case:	Stainless steel, Ø 40 mm, height 12.1 mm; bezel with ceramic insert, bidirectional rotation, 24-hour graduation, sapphire glass; screw-down crown; water-resistant to 10 bar
Bracelet:	Oyster stainless steel, folding Oysterlock safety clasp with extension link
Price:	$9700 (€ 7200)

TECHNICAL DATA

OYSTER PERPETUAL GMT-MASTER II

Reference: 116710 LN

Movement: Automatic, Rolex caliber 3186; Ø 28.5 mm, height 6.4 mm; 31 jewels; 28,800 A/h; certified Chronometer (COSC)

Functions: Hours, minutes, center seconds; additional 24-hour display; date

Case: Stainless steel, Ø 40 mm, height 12.1 mm; bezel with ceramic insert, bidirectional rotation, 24-hour graduation, sapphire glass; screw-down crown; water-resistant to 10 bar

Bracelet: Oyster, stainless steel, folding Oysterlock safety clasp with

OYSTER PERPETUAL GMT-MASTER II

Reference: 116759 SA

Movement: Automatic, Rolex caliber 3186; Ø 28.5 mm, height 6.4 mm; 31 jewels; 28,800 A/h; certified chronometer (COSC)

Functions: Hours, minutes, center seconds; additional 24-hour display; date

Case: Stainless steel, Ø 40 mm, height 12.1 mm; bezel set with 30 sapphires and 29 diamonds, bidirectional rotation, sapphire glass; screw-down crown; water-resistant to 10 bar

Bracelet: Oyster, stainless steel and yellow gold, folding Oysterlock

OYSTER PERPETUAL GMT-MASTER II

Reference: 116713LN
Movement: Automatic, Rolex caliber 3186; Ø 28.5 mm, height 6.4 mm; 31 jewels; 28,800 A/h; certified Chronometer (COSC)
Functions: Hours, minutes, center seconds; additional 24-hour display; date
Case: Stainless steel, Ø 40 mm, height 12.1 mm; bezel in yellow gold with ceramic insert, bidirectional rotation, 24-hour graduation, sapphire glass; screw-down crown; water-resistant to 10 bar
Bracelet: Oyster, stainless steel and yellow gold, folding Oysterlock safety clasp with extension link
Price: $14,100 (€ 10,500)

OYSTER PERPETUAL GMT-MASTER II

Reference: 116718 LN
Movement: Automatic, Rolex caliber 3186; Ø 28.5 mm, height 6.4 mm; 31 jewels; 28,800 A/h; certified chronometer (COSC)
Functions: Hours, minutes, center seconds; additional 24-hour display; date
Case: Yellow gold, Ø 40 mm, height 12.1 mm; bezel in yellow gold with ceramic insert, bidirectional rotation, 24-hour graduation, sapphire glass; screw-down crown; water-resistant to 10 bar
Bracelet: Oyster, yellow gold, folding Oysterlock safety clasp with extension link
Price: $36,100 (€ 26,900)

ROLEX
SKY-DWELLER

In 2012, Rolex introduced the Sky-Dweller, a model that caters to the demands of the modern globetrotter. In addition to the current time in your respective location, home time can always be kept in view too. An inconspicuous calendar that displays the day and month is also integrated. Different functions can be accessed using the three different positions of the Ring Command bezel, which are then comfortably set with the crown: the reference time on the 24-hour disc, the date under Rolex's typical Cyclops lens, and the months ,which are displayed in apertures over the hour indexes. Rolex developed the caliber 9001 movement specifically for the Sky-Dweller; it blends precision and dependability with unique complication functions for the Swiss brand.

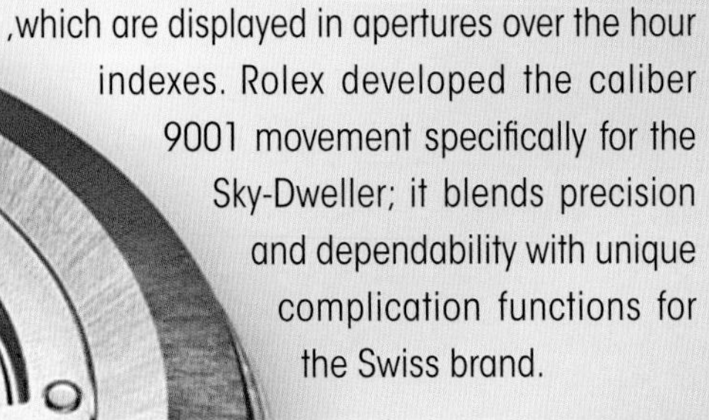

SKY-DWELLER

Reference: 326939

Movement: Automatic, Rolex caliber 9001; Ø 33 mm, height 8 mm; 40 jewels; 28,800 A/h; 72-hour power reserve; Parachrom hairspring; Glucydur balance with Microstella regulating screws; certified chronometer (COSC)

Functions: Hours, minutes, center seconds; second time zone (additional 24-hour display); annual calendar with date and month

Case: White gold, Ø 42 mm, height 14.1 mm; bezel bidirectional rotation for controlling functions; sapphire glass; lens over date aperture; screw-down crown; water-resistant to 10 bar

Bracelet: Oyster, white gold, folding Oysterclasp

Price: $53,000 (€ 39,500)

TECHNICAL DATA

SKY-DWELLER

Reference:	326938
Movement:	Automatic, Rolex caliber 9001; Ø 33 mm, height 8 mm; 40 jewels; 28,800 A/h; 72-hour power reserve; Parachrom hairspring; Glucydur balance with Microstella regulating screws; certified chronometer (COSC)
Functions:	Hours, minutes, center seconds; second time zone (additional 24-hour display); annual calendar with date and month
Case:	Yellow gold, Ø 42 mm, height 14.1 mm; bezel bidirectional rotation for controlling functions; sapphire glass; lens over date aperture; screw-down crown; water-resistant to 10 bar
Bracelet:	Oyster, Yellow gold, folding Oysterclasp
Price:	$50,000 (€ 37,300)

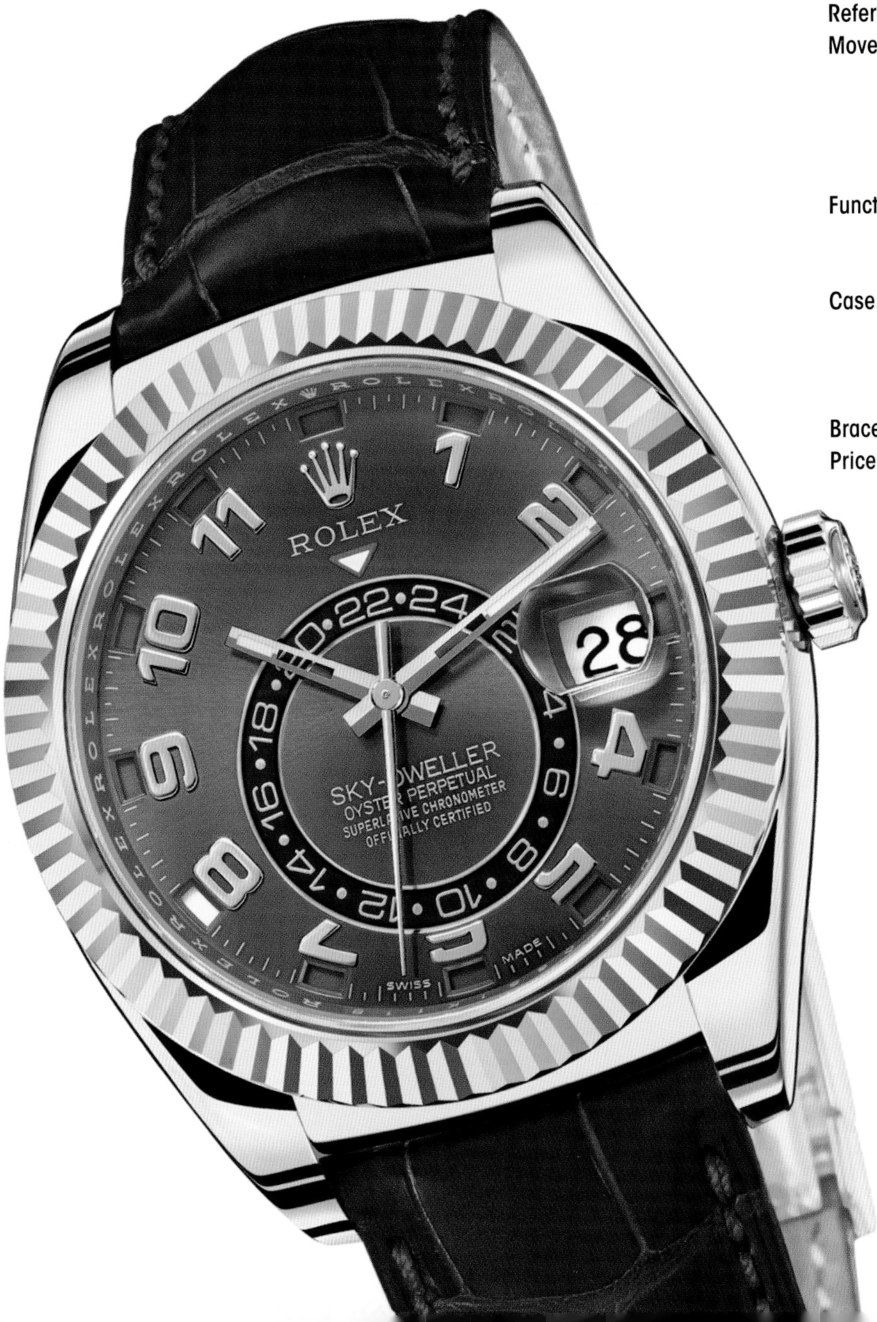

SKY-DWELLER

Reference:	326135
Movement:	Automatic, Rolex caliber 9001; Ø 33 mm, height 8 mm; 40 jewels; 28,800 A/h; 72-hour power reserve; Parachrom hairspring; Glucydur balance with Microstella regulating screws; certified chronometer (COSC)
Functions:	Hours, minutes, center seconds; second time zone (additional 24-hour display); annual calendar with date and month
Case:	Rose gold, Ø 42 mm, height 14.1 mm; bezel bidirectional rotation for controlling functions; sapphire glass; lens over date aperture; screw-down crown; water-resistant to 10 bar
Bracelet:	Reptile leather, folding Oysterclasp
Price:	$42,800 (€ 31,900)

ROLEX SUBMARINER / SUBMARINER DATE

The Submariner was first launched in 1953, and it is continuously being improved. This mother of all diver's watches has seen countless imitators in the last few decades, but few can come close to matching it. With its robust Oyster construction, which guarantees absolute water-resistance even at greater depths, the easy-to-use bezel for setting diving times and its perfect readability underwater even at twilight, the Submariner has set the standard. Since 1969, it is also available with date display, making it a watch that is no longer just for divers, but also for everyday use or for formal dress.

OYSTER PERPETUAL SUBMARINER

Reference: 114060

Movement: Automatic, Rolex caliber 3130; Ø 28.5 mm, height 5.85 mm; 31 jewels; 28,800 A/h; Parachrom hairspring, Glucydur balance with Microstella regulating screws; 48-hour power reserve; certified chronometer (COSC)

Functions: Hours, minutes, center seconds

Case: Stainless steel, Ø 40 mm, height 12.5 mm; bezel with ceramic insert, unidirectional rotation, 60-minute graduation, sapphire glass; screw-down crown; water-resistant to 30 bar

Bracelet: Oyster, Stainless steel, folding Oysterlock safety clasp with extension link

Price: $8000 (€ 6000)

TECHNICAL DATA

OYSTER PERPETUAL SUBMARINER DATE

Reference:	116610LN
Movement:	Automatic, Rolex caliber 3135; Ø 28.5 mm, height 6 mm; 31 jewels; 28,800 A/h; Parachrom hairspring, Glucydur balance with Microstella regulating screws; 48-hour power reserve; certified chronometer (COSC)
Functions:	Hours, minutes, center seconds; date
Case:	Stainless steel, Ø 40 mm, height 12.5 mm; bezel with ceramic insert, unidirectional rotation, 60-minute graduation, sapphire glass; screw-down crown; water-resistant to 30 bar
Bracelet:	Oyster, stainless steel, folding Oysterlock safety clasp
Price:	$9200 (€ 6850)

OYSTER PERPETUAL SUBMARINER DATE

Reference:	116613LB
Movement:	Automatic, Rolex caliber 3135; Ø 28.5 mm, height 6 mm; 31 jewels; 28,800 A/h; Parachrom hairspring, Glucydur balance with Microstella regulating screws; 48-hour power reserve; certified chronometer (COSC)
Functions:	Hours, minutes, center seconds; date
Case:	Stainless steel, Ø 40 mm, height 12.5 mm; bezel made of yellow gold with ceramic insert, unidirectional rotation, 60-minute graduation, sapphire glass; screw-down crown; water-resistant to 30 bar
Bracelet:	Oyster, stainless steel and yellow gold, folding Oysterlock safety clasp
Price:	$14,600 (€ 10,850)

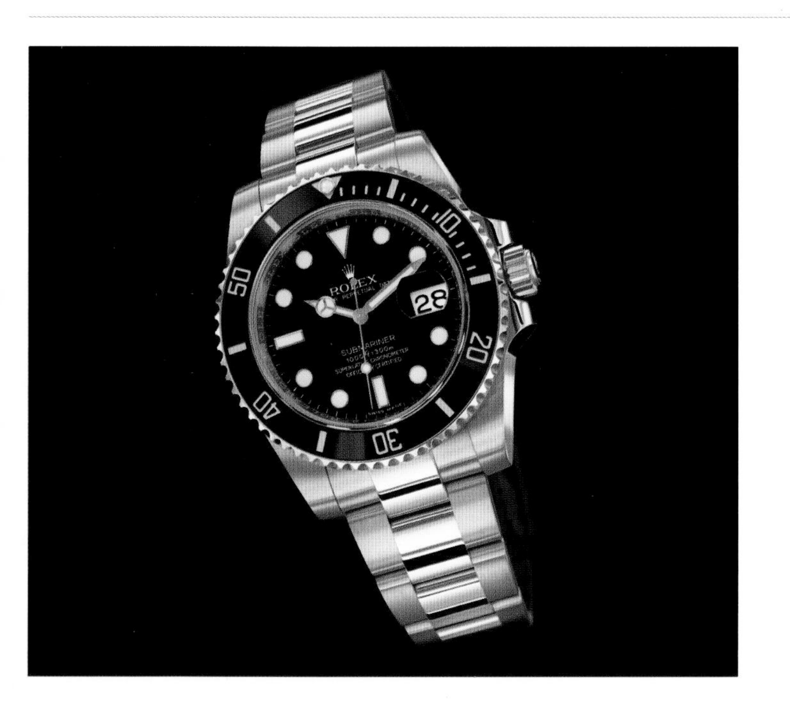

OYSTER PERPETUAL SUBMARINER DATE

Reference: 116618 LN
Movement: Automatic, Rolex caliber 3135; Ø 28.5 mm, height 6 mm; 31 jewels; 28,800 A/h; Parachrom hairspring, Glucydur balance with Microstella regulating screws; 48-hour power reserve; certified chronometer (COSC)
Functions: Hours, minutes, center seconds; date
Case: Yellow gold, Ø 40 mm, height 12.5 mm; bezel with ceramic insert, unidirectional rotation, 60- minute graduation, sapphire glass; screw-down crown; water-resistant to 30 bar
Bracelet: Yellow gold, folding Oysterlock safety clasp
Price: $37,000 (€ 27,600)

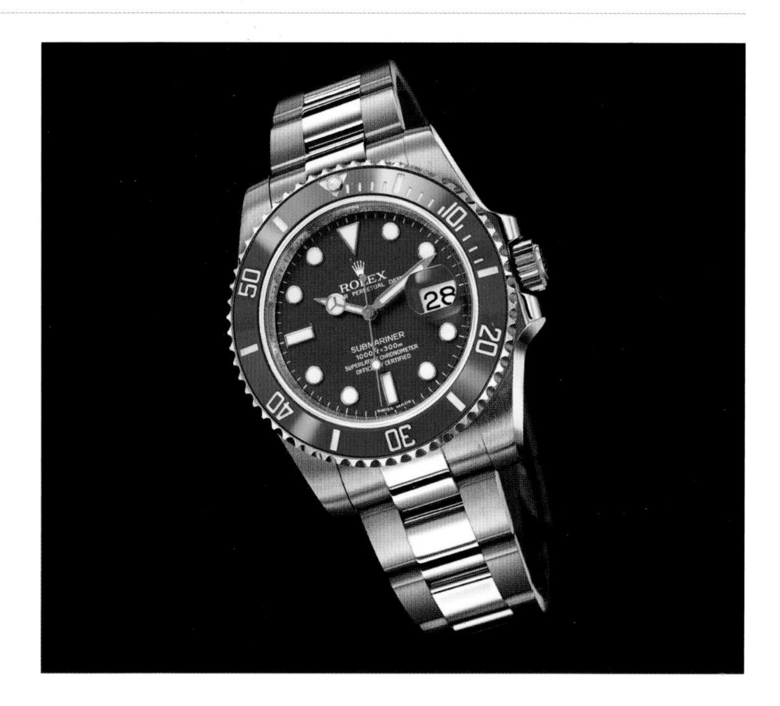

OYSTER PERPETUAL SUBMARINER DATE

Reference: 116619LB
Movement: Automatic, Rolex caliber 3135; Ø 28.5 mm, height 6 mm; 31 jewels; 28,800 A/h; Parachrom hairspring, Glucydur balance with Microstella regulating screws; 48-hour power reserve; certified chronometer (COSC)
Functions: Hours, minutes, center seconds; date
Case: White gold, Ø 40 mm, height 12.5 mm; bezel with ceramic insert, unidirectional rotation, 60-minute graduation, sapphire glass; screw-down crown; water-resistant to 30 bar
Bracelet: White gold, folding Oysterlock safety clasp
Price: $40,000 (€ 29,800)

ROLEX DEEPSEA

While the Sky-Dweller prowls the skies, the Sea-Dweller's cousin, the Deepsea, dives down to the deepest depths of the oceans. The Deepsea was re-introduced in 2008, and is probably the most durable diver's watch that Rolex has ever made. It is a high-performance diver's watch capable of reaching depths where very few living things exist. The stainless steel body of the watch is furnished with the Ringlock system and can withstand up to 390 bar of water pressure. This equates to the pressure at a depth of 3,900 meters. Extra-thick sapphire glass, titanium case back, and the Triplock crown with its three sealing systems round out the Oyster construction. The especially large indexes and hands are coated with blue luminescent material to guarantee optimal readability underwater.

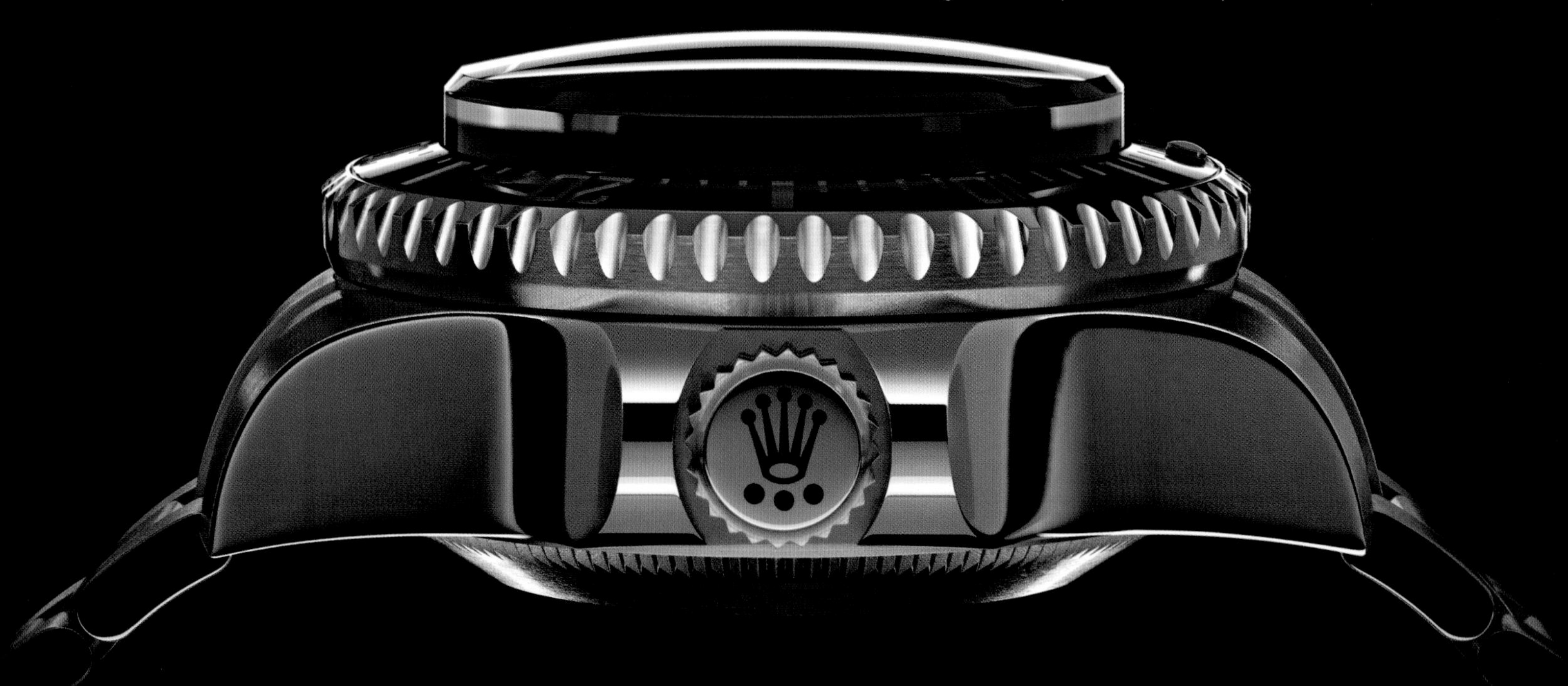

TECHNICAL DATA

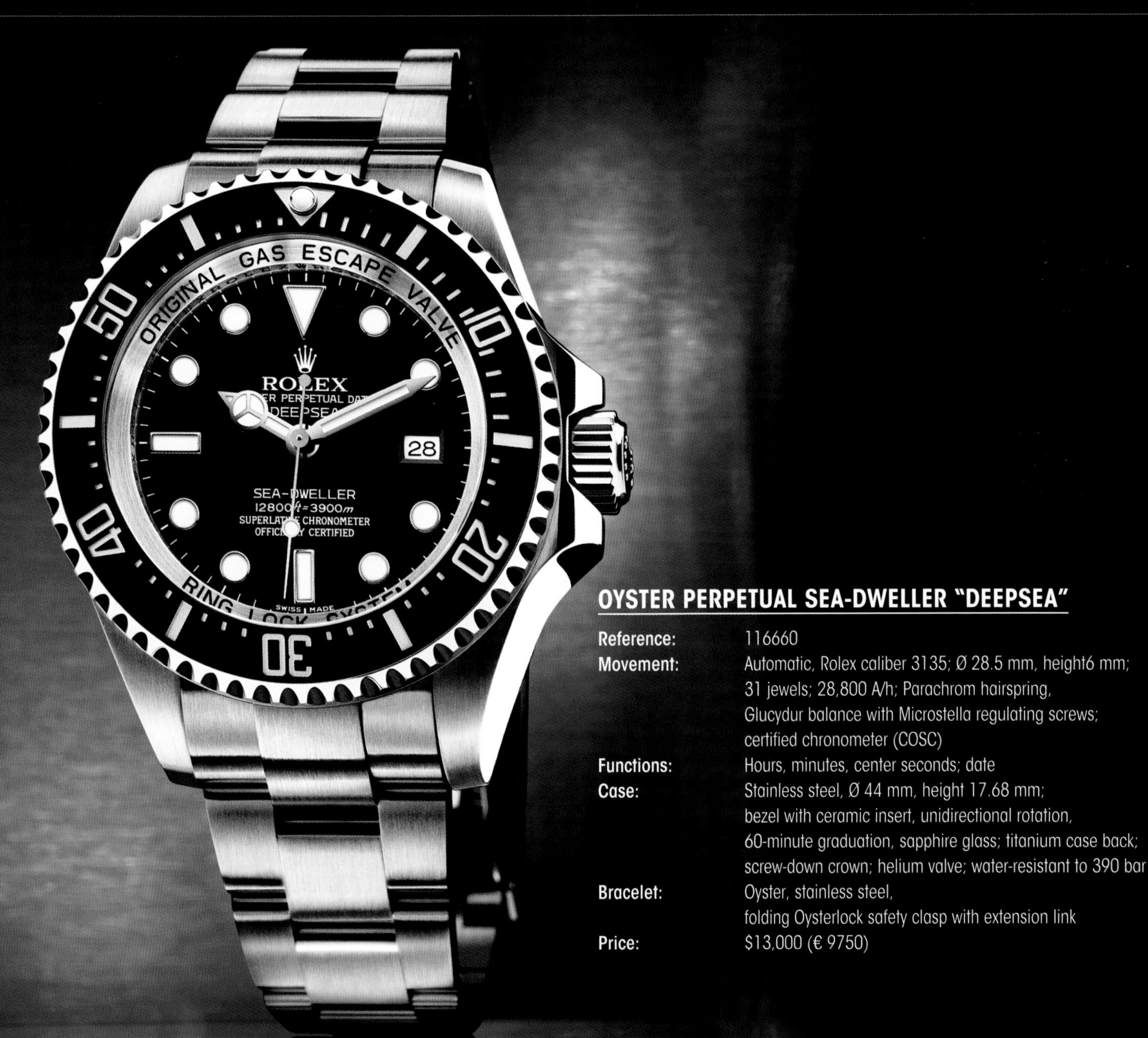

OYSTER PERPETUAL SEA-DWELLER "DEEPSEA"

Reference: 116660

Movement: Automatic, Rolex caliber 3135; Ø 28.5 mm, height6 mm; 31 jewels; 28,800 A/h; Parachrom hairspring, Glucydur balance with Microstella regulating screws; certified chronometer (COSC)

Functions: Hours, minutes, center seconds; date

Case: Stainless steel, Ø 44 mm, height 17.68 mm; bezel with ceramic insert, unidirectional rotation, 60-minute graduation, sapphire glass; titanium case back; screw-down crown; helium valve; water-resistant to 390 bar

Bracelet: Oyster, stainless steel, folding Oysterlock safety clasp with extension link

Price: $13,000 (€ 9750)

ROLEX COSMOGRAPH DAYTONA

First launched in 1963, the Cosmograph Daytona (named after the legendary race track in Daytona Beach, Florida) is now among the most coveted and famous watches in the Rolex collection. Early models were not a big success due to their hand-wound movements, which were then considered behind the times. But this has changed radically in the last few years. The movements, which had previously been purchased from third-party manufacturers, have been made in-house since 2000. From its inception, the stability, precision, and dependability of the chronograph caliber 4130 was impressive, and today it is considered the best chronograph movement currently available on the market. While there are dozens of versions of the modern Cosmograph Daytona for every taste and market, the early models—once notorious shelf warmers—are now popular objects for collectors.

TECHNICAL DATA

OYSTER PERPETUAL COSMOGRAPH DAYTONA

Reference: 116506

Movement: Automatic, Rolex caliber 4130; Ø 30.5 mm, height 6.5 mm; 44 jewels; 28,800 A/h; Parachrom hairspring; certified chronometer (COSC)

Functions: Hours, minutes, auxiliary seconds; chronograph

Case: platinum, Ø 40 mm, height 12.1 mm; ceramic bezel, tachymeter scale, sapphire glass; screw-down crown and pushers; water-resistant to 10 bar

Bracelet: President, platinum, folding Oysterlock safety clasp

Price: $81,400 (€ 60,650)

TECHNICAL DATA

OYSTER PERPETUAL COSMOGRAPH DAYTONA

Reference:	116515LN
Movement:	Automatic, Rolex caliber 4130; Ø 30.5 mm, height 6.5 mm; 44 jewels; 28,800 A/h; Parachrom hairspring; certified chronometer (COSC)
Functions:	Hours, minutes, auxiliary seconds; chronograph
Case:	Rose gold, Ø 40 mm, height 12.1 mm; ceramic bezel, tachymeter scale, sapphire glass; screw-down crown and pushers; water-resistant to 10 bar
Bracelet:	Reptile leather, folding Oysterlock safety clasp
Price:	$31,200 (€ 23,250)

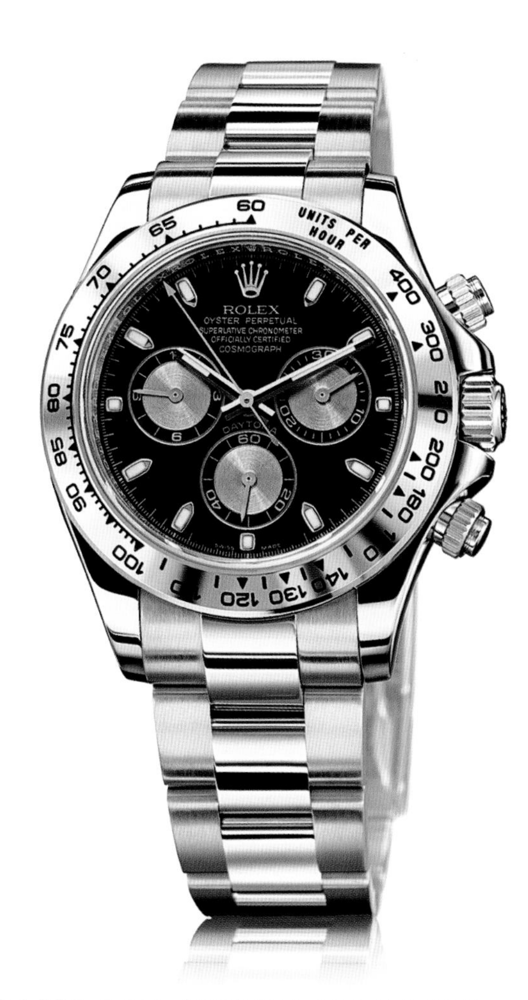

OYSTER PERPETUAL COSMOGRAPH DAYTONA

Reference: 116505
Movement: Automatic, Rolex caliber 4130; Ø 30.5 mm, height 6.5 mm; 44 jewels; 28,800 A/h; Parachrom hairspring; certified chronometer (COSC)
Functions: Hours, minutes, auxiliary seconds; chronograph
Case: Rose gold, Ø 40 mm, height 12.8 mm; bezel with tachymeter scale, sapphire glass; screw-down crown and pushers; water-resistant to 10 bar
Bracelet: Oyster, rose gold, folding Oysterlock safety clasp with extension link
Price: $40,700 (€ 30,300)

OYSTER PERPETUAL COSMOGRAPH DAYTONA

Reference: 116518
Movement: Automatic, Rolex caliber 4130; Ø 30.5 mm, height 6.5 mm; 44 jewels; 28,800 A/h; Parachrom hairspring; certified chronometer (COSC)
Functions: Hours, minutes, auxiliary seconds; chronograph
Case: Yellow gold, Ø 40 mm, height 12.8 mm; bezel with tachymeter scale, sapphire glass; screw-down crown and pushers; water-resistant to 10 bar
Bracelet: Reptile leather, folding Oysterlock safety clasp
Price: $27,300 (€ 20,350)

TECHNICAL DATA

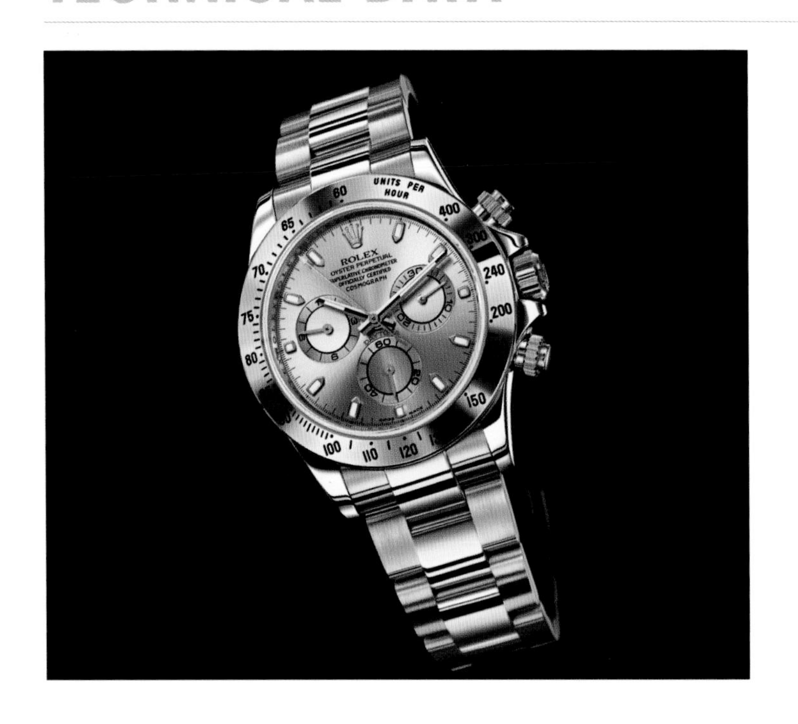

OYSTER PERPETUAL COSMOGRAPH DAYTONA

Reference: 116523
Movement: Automatic, Rolex caliber 4130; Ø 30.5 mm, height 6.5 mm; 44 jewels; 28,800 A/h; Parachrom hairspring; certified Chronometer (COSC)
Functions: Hours, minutes, auxiliary seconds; chronograph
Case: Stainless steel, Ø 40 mm, height 12.8 mm; bezel in yellow gold with tachymeter scale, sapphire glass; screw-down crown and pushers; water-resistant to 10 bar
Bracelet: Oyster, stainless steel and yellow gold, folding Oysterlock safety clasp
Price: $18,300 (€ 13,650)

OYSTER PERPETUAL COSMOGRAPH DAYTONA

Reference: 116509
Movement: Automatic, Rolex caliber 4130; Ø 30.5 mm, height 6.5 mm; 44 jewels; 28,800 A/h; Parachrom hairspring; certified chronometer (COSC)
Functions: Hours, minutes, auxiliary seconds; chronograph
Case: White gold, Ø 40 mm, height 12.8 mm; bezel with tachymeter scale, sapphire glass; screw-down crown and pushers; water-resistant to 10 bar
Bracelet: Reptile leather, folding Oysterlock safety clasp
Price: $40,700 (€ 30,300)

OYSTER PERPETUAL COSMOGRAPH DAYTONA

Reference:	116528
Movement:	Automatic, Rolex caliber 4130; Ø 30.5 mm, height 6.5 mm; 44 jewels; 28,800 A/h; Parachrom hairspring; certified Chronometer (COSC)
Functions:	Hours, minutes, auxiliary seconds; chronograph
Case:	Yellow gold, Ø 40 mm, height 12.8 mm; bezel with tachymeter scale, sapphire glass; screw-down crown and pushers; water-resistant to 10 bar
Bracelet:	Oyster, yellow gold, folding Oysterlock safety clasp
Price:	$37,600 (€ 28,000)

ROLEX
YACHT-MASTER

The Yacht-Master series documents Rolex's ties to professional sailing. It combines the ruggedness and look of a sport watch with a functionality that can decide victory or defeat aboard a sail boat. The Yacht-Master II's special chronograph also has a mechanical complication that was specifically tailored to the needs of this sport: the countdown function, which can be programmed to the countdown time in a regatta and be retrieved over and over again if needed. For setting, the Rolex-designed Ring Command bezel, which works in conjunction with the start/stop and reset pushers, ensures that the countdown can be activated precisely.

OYSTER PERPETUAL YACHT-MASTER

Reference: 116622
Movement: Automatic, Rolex caliber 3135; Ø 28.5 mm, height 6 mm; 31 jewels; 28,800 A/h; Parachrom hairspring; ca. 72-hour power reserve; certified chronometer (COSC)
Functions: Hours, minutes, center seconds; date
Case: Stainless steel, Ø 40 mm, height 11.7 mm; bezel in platinum, bidirectional rotation, with 60-minute graduation, sapphire glass; screw-down crown; water-resistant to 10 bar
Bracelet: Oyster, stainless steel, folding Oysterlock safety clasp with extension link
Price: $12,600 (€ 9350)

TECHNICAL DATA

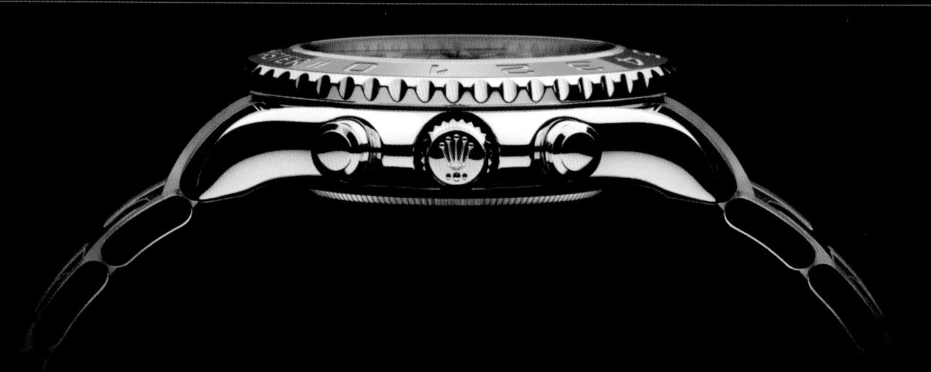

OYSTER PERPETUAL YACHT-MASTER II REGATTA CHRONOGRAPH

Reference: 116681

Movement: Automatic, Rolex caliber 4160 (Basis Rolex caliber 4130); Ø 31.2 mm, height 8.05 mm; 42 jewels; 28,800 A/h; Parachrom hairspring; ca. 72-hour power reserve; certified chronometer (COSC)

Functions: Hours, minutes, auxiliary seconds; programmable regatta countdown with mechanical memory

Case: Stainless steel, Ø 44 mm, height 13.8 mm; bezel in rose gold with ceramic insert, bidirectional rotation, sapphire glass; screw-down crown; water-resistant to 10 bar

Bracelet: Oyster, stainless steel and rose gold, folding Oysterlock safety clasp with extension link

Price: $27,200 (€ 20,300)

TECHNICAL DATA

OYSTER PERPETUAL YACHT-MASTER II REGATTA CHRONOGRAPH

Reference:	116688
Movement:	Automatic, Rolex caliber 4160 (Basis Rolex caliber 4130); Ø 31.2 mm, height 8.05 mm; 42 jewels; 28,800 A/h; Parachrom hairspring; ca. 72-hour power reserve; certified chronometer (COSC)
Functions:	Hours, minutes, auxiliary seconds; programmable regatta countdown with mechanical memory
Case:	Yellow gold, Ø 44 mm, height 13.8 mm; bezel with ceramic insert, bidirectional rotation, sapphire glass; screw-down crown; water-resistant to 10 bar
Bracelet:	Oyster, yellow gold, folding Oysterlock safety clasp with extension link
Price:	$27,300 (€ 35,200)

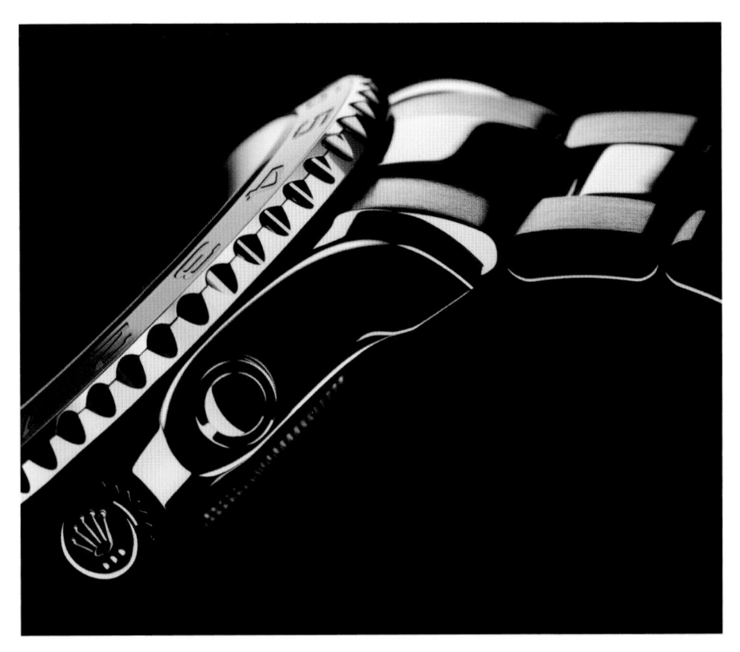

TECHNICAL DATA

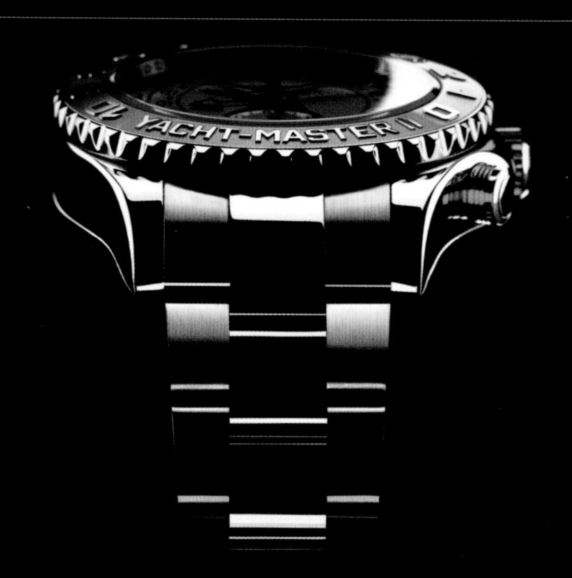

OYSTER PERPETUAL YACHT-MASTER II REGATTA CHRONOGRAPH

Reference:	116689
Movement:	Automatic, Rolex caliber 4160 (Basis Rolex caliber 4130); Ø 31.2 mm, height 8.05 mm; 42 jewels; 28,800 A/h; Parachrom hairspring; ca. 72-hour power reserve; certified chronometer (COSC)
Functions:	Hours, minutes, auxiliary seconds; programmable regatta countdown with mechanical memory
Case:	White gold, Ø 44 mm, height 13.8 mm; bezel in rose gold with ceramic insert, bidirectional rotation, sapphire glass; screw-down crown; water-resistant to 10 bar
Bracelet:	Oyster, white gold, folding Oysterlock safety clasp with extension link
Price:	$52,200 (€ 38,900)

OYSTER PERPETUAL YACHT-MASTER II REGATTA CHRONOGRAPH

Reference: 116680

Movement: Automatic, Rolex caliber 4160 (Basis Rolex caliber 4130); Ø 31.2 mm, height 8.05 mm; 42 jewels; 28,800 A/h; Parachrom hairspring; ca. 72-hour power reserve; certified chronometer (COSC)

Functions: Hours, minutes, auxiliary seconds; programmable regatta countdown with mechanical memory

Case: Stainless steel, Ø 44 mm, height 13.8 mm; bezel in rose gold with ceramic insert, bidirectional rotation, sapphire glass; screw-down crown; water-resistant to 10 bar

Bracelet: Oyster, stainless steel, folding Oysterlock safety clasp with extension link

Price: $20,200 (€ 15,050)

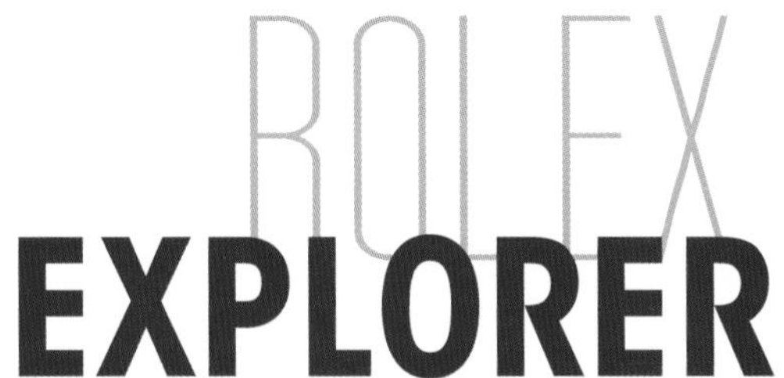

ROLEX EXPLORER

On the roof of the world with Rolex: In 1953, after several failed attempts, Sir Edmund Hillary and his Nepalese partner Tenzig Norgay made it to the top of Mount Everest, the highest mountain in the world at 8,848 meters. On his wrist he wore a precursor to the Explorer from Rolex, a watch that could withstand the extreme conditions of cold and altitude. The modern Explorer proudly reflects this significant legacy. The case, made from solid block stainless steel, resists shocks and impacts and is particularly resistant to corrosion. Chromalight luminescence on hands and indexes ensures perfect readability in poor lighting conditions. Explorer II, its somewhat bigger brother with a case diameter of 42 millimeters, even has a date function and a simultaneous 24-hour display so that the overall view is never lost on research expeditions the world over.

OYSTER PERPETUAL EXPLORER

Reference: 214270

Movement: Automatic, Rolex caliber 3132 (Basis Rolex caliber 3135); Ø 30.97 mm, height 5.87 mm; 31 jewels; 28,800 A/h; Parachrom hairspring, Paraflex shock absorbers; certified chronometer (COSC)

Functions: Hours, minutes, center seconds

Case: Stainless steel, Ø 39 mm, height 11.9 mm; sapphire glass; screw-down crown; water-resistant to 10 bar

Bracelet: Oyster, stainless steel, folding Oysterlock safety clasp with extension link

Price: $7000 (€ 5250)

TECHNICAL DATA

OYSTER PERPETUAL EXPLORER II

Reference:	216570
Movement:	Automatic, Rolex caliber 3187 (Basis Rolex caliber 3135); Ø 28.5 mm, height 6.4 mm; 31 jewels; 28,800 A/h; Parachrom hairspring, Paraflex shock absorbers; certified chronometer (COSC)
Functions:	Hours, minutes, center seconds; additional 24-hour display; date
Case:	Stainless steel, Ø 42 mm, height 12.3 mm; bezel with 24-hour graduation, sapphire glass; screw-down crown; water-resistant to 10 bar
Bracelet:	Oyster, stainless steel, folding Oysterlock safety clasp with extension link
Price:	$8700 (€ 6500)

TECHNICAL DATA

OYSTER PERPETUAL EXPLORER II

Reference:	216570
Movement:	Automatic, Rolex caliber 3187 (Basis Rolex caliber 3135); Ø 28.5 mm, height 6.4 mm; 31 jewels; 28,800 A/h; Parachrom hairspring, Paraflex shock absorbers; certified chronometer (COSC)
Functions:	Hours, minutes, center seconds; additional 24-hour display; date
Case:	Stainless steel, Ø 42 mm, height 12.3 mm; bezel with 24-hour graduation, sapphire glass; screw-down crown; water-resistant to 10 bar
Bracelet:	Oyster, stainless steel, folding Oysterlock safety clasp with extension link
Price:	$8700 (€ 6500)

DATEJUST

In 1945, shortly after World War II, Rolex introduced the Oyster Perpetual Datejust. Based on the Oyster collection, it represented the first automatic, water-resistant watch with a chronometer certificate, and it also displayed the date on the dial. Originally intended as a simple men's watch, it appeared on the market in a variety of versions over the following decades and today is an elegant alternative to the models of the various professional lines. Meanwhile, the innovations that are typical for Rolex regarding its movement escapement have also found their way into the Datejust, which guarantees chronometer quality as well as absolute durability and dependability. The Datejust II, in contrast to the original model, has the optimized caliber 3136 and a case diameter of 41 millimeters.

OYSTER PERPETUAL DATEJUST

Reference: 116200
Movement: Automatic, Rolex caliber 3135; Ø 28.5 mm, height 6 mm; 31 jewels; 28,800 A/h; Breguet spiral, Glucydur balance with Microstella regulating screws; certified chronometer (COSC)
Functions: Hours, minutes, center seconds; date
Case: Stainless steel, Ø 36 mm, height 11.8 mm; sapphire glass; screw-down crown; water-resistant to 10 bar
Bracelet: Oyster, stainless steel, folding clasp with extension link
Price: $7200 (€ 5350)

TECHNICAL DATA

OYSTER PERPETUAL DATEJUST

Reference: 116233
Movement: Automatic, Rolex caliber 3135; Ø 28.5 mm, height 6 mm; 31 jewels; 28,800 A/h; Breguet spiral, Glucydur balance with Microstella regulating screws; certified chronometer (COSC)
Functions: Hours, minutes, center seconds; date
Case: Stainless steel, Ø 36 mm, height 11.8 mm; bezel in yellow gold; sapphire glass; screw-down crown; water-resistant to 10 bar
Bracelet: Jubilee, stainless steel and yellow gold, folding clasp with extension link
Price: $11, 800 (€ 8800)

OYSTER PERPETUAL DATEJUST

Reference: 116233
Movement: Automatic, Rolex caliber 3135; Ø 28.5 mm, height 6 mm; 31 jewels; 28,800 A/h; Breguet spiral, Glucydur balance with Microstella regulating screws; certified Chronometer (COSC)
Functions: Hours, minutes, center seconds; date
Case: Stainless steel, Ø 36 mm, height 11.8 mm; bezel in yellow gold; sapphire glass; screw-down crown; water-resistant to 10 bar
Bracelet: Jubilee, stainless steel and yellow gold, folding clasp with extension link
Price: $11, 800 (€ 8800)

TECHNICAL DATA

OYSTER PERPETUAL DATEJUST

Reference: 116231
Movement: Automatic, Rolex caliber 3135; Ø 28.5 mm, height 6 mm; 31 jewels; 28,800 A/h; Breguet spiral, Glucydur balance with Microstella regulating screws; certified chronometer (COSC)
Functions: Hours, minutes, center seconds; date
Case: Stainless steel, Ø 36 mm, height 11.8 mm; bezel in rose gold; sapphire glass; screw-down crown; water-resistant to 10 bar
Bracelet: Jubilee, stainless steel and rose gold, folding clasp with extension link
Price: $8700 (€ 6500)

OYSTER PERPETUAL DATEJUST

Reference: 116234
Movement: Automatic, Rolex caliber 3135; Ø 28.5 mm, height 6 mm; 31 jewels; 28,800 A/h; Breguet spiral, Glucydur balance with Microstella regulating screws; certified chronometer (COSC)
Functions: Hours, minutes, center seconds; date
Case: Stainless steel, Ø 36 mm, height 11.8 mm; bezel in white gold; sapphire glass; screw-down crown; water-resistant to 10 bar
Bracelet: Oyster, stainless steel, folding clasp with extension link
Price: $8525 (€ 6350)

TECHNICAL DATA

OYSTER PERPETUAL DATEJUST II

Reference: 116300
Movement: Automatic, Rolex caliber 3136; 30.97 mm, height 6.47 mm; 31 jewels; 28,800 A/h; Parachrom hairspring, Paraflex shock absorbers; certified chronometer (COSC)
Functions: Hours, minutes, center seconds; date
Case: Stainless steel, Ø 41 mm, height 11.8 mm; sapphire glass; screw-down crown; water-resistant to 10 bar
Bracelet: Oyster, stainless steel, folding Oysterclasp with extension link
Price: $7700 (€ 5750)

TECHNICAL DATA

OYSTER PERPETUAL DATEJUST II

Reference: 116334

Movement: Automatic, Rolex caliber 3136; Ø 30.97 mm, height 6.47 mm; 31 jewels; 28,800 A/h; Parachrom hairspring, Paraflex shock absorbers; certified Chronometer (COSC)

Functions: Hours, minutes, center seconds; date

Case: Stainless steel, Ø 41 mm, height 11.8 mm; bezel in white gold; sapphire glass; screw-down crown; water-resistant to 10 bar

Bracelet: Oyster, stainless steel, folding Oysterclasp with extension link

Price: $9900 (€ 7400)

OYSTER PERPETUAL DATEJUST II

Reference: 116333

Movement: Automatic, Rolex caliber 3136; Ø 30.97 mm, height 6.47 mm; 31 jewels; 28,800 A/h; Parachrom hairspring, Paraflex shock absorbers; certified chronometer (COSC)

Functions: Hours, minutes, center seconds; date

Case: Stainless steel, Ø 41 mm, height 11.8 mm; bezel in yellow gold; sapphire glass; screw-down crown; water-resistant to 10 bar

Bracelet: Oyster, Stainless steel and yellow gold, folding Oysterclasp with extension link

Price: $12,600 (€ 9400)

ROLEX
DAY-DATE

In 1956, Rolex presented yet another of its many classics from the fifties, which still shape its collection today. Only available in gold or platinum, the Day-Date was the first wristwatch to display the fully-written weekday in a separate window next to the date. It was worn primarily by well-known and influential personalities and was quickly regarded as the "President's watch." To improve readability, a section of sapphire glass with two-and-a-half times magnification above the date display was integrated, the so-called Cyclops lens. Little by little, this innovation was installed on all Rolex watches with date displays and it became a distinctive feature of the brand. In 2008, the Day-Date II was equipped with the new caliber 3156 and enlarged to a 41 millimeter case.

OYSTER PERPETUAL DAY-DATE

Reference: 118206
Movement: Automatic, Rolex caliber 3155 (Basis Rolex caliber 3135); Ø 28.5 mm, height 6.45 mm; 31 jewels; 28,800 A/h; Glucydur balance with Microstella regulating screws; certified chronometer (COSC)
Functions: Hours, minutes, center seconds; date and weekday
Case: Platinum, Ø 36 mm, height 11.9 mm; sapphire glass; screw-down crown; water-resistant to 10 bar
Bracelet: President, platinum, folding clasp
Price: $62,400 (€ 46,550)

TECHNICAL DATA

OYSTER PERPETUAL DAY-DATE

Reference:	118138
Movement:	Automatic, Rolex caliber 3155 (Basis Rolex caliber 3135); Ø 28.5 mm, height 6.45 mm; 31 jewels; 28,800 A/h; Glucydur balance with Microstella regulating screws; certified chronometer (COSC)
Functions:	Hours, minutes, center seconds; date and weekday
Case:	Yellow gold, Ø 36 mm, height 11.9 mm; sapphire glass; screw-down crown; water-resistant to 10 bar
Bracelet:	Reptile leather, folding clasp
Price:	$24,000 (€ 17,850)

OYSTER PERPETUAL DAY-DATE

Reference: 118139

Movement: Automatic, Rolex caliber 3155 (Basis Rolex caliber 3135); Ø 28.5 mm, height 6.45 mm; 31 jewels; 28,800 A/h; Glucydur balance with Microstella regulating screws; certified chronometer (COSC)

Functions: Hours, minutes, center seconds; date and weekday

Case: White gold, Ø 36 mm, height 11.9 mm; sapphire glass; screw-down crown; water-resistant to 10 bar

Bracelet: Reptile leather, folding clasp

Price: $25,600 (€ 19,050)

OYSTER PERPETUAL DAY-DATE

Reference: 118135

Movement: Automatic, Rolex caliber 3155 (Basis Rolex caliber 3135); Ø 28.5 mm, height 6.45 mm; 31 jewels; 28,800 A/h; Glucydur balance with Microstella regulating screws; certified Chronometer (COSC)

Functions: Hours, minutes, center seconds; date und weekday

Case: Rose gold, Ø 36 mm, height 11.9 mm; sapphire glass; screw-down crown; water-resistant to 10 bar

Bracelet: Reptile leather, folding clasp

Price: $25,600 (€ 19,050)

TECHNICAL DATA

OYSTER PERPETUAL DAY-DATE II

Reference: 218206

Movement: Automatic, Rolex caliber 3156 (Basis Rolex caliber 3135); Ø 30.97 mm, height 6.47 mm; 31 jewels; 28,800 A/h; Parachrom hairspring, Paraflex shock absorbers, Glucydur balance with Microstella regulating screws; certified chronometer (COSC)

Functions: Hours, minutes, center seconds; date and weekday

Case: Platinum, Ø 41 mm, height 11.9 mm; sapphire glass; screw-down crown; water-resistant to 10 bar

Bracelet: President, platinum, folding clasp

Price: $67,700 (€ 50,450)

OYSTER PERPETUAL DAY-DATE II

Reference: 218235

Movement: Automatic, Rolex caliber 3156 (Basis Rolex caliber 3135); Ø 30.97 mm, height 6.47 mm; 31 jewels; 28,800 A/h; Parachrom hairspring, Paraflex shock absorbers, Glucydur balance with Microstella regulating screws; certified chronometer (COSC)

Functions: Hours, minutes, center seconds; date and weekday

Case: Rose gold, Ø 41 mm, height 11.9 mm; sapphire glass; screw-down crown; water-resistant to 10 bar

Bracelet: President, rose gold, folding clasp

Price: $40,700 (€ 30,300)

TECHNICAL DATA

OYSTER PERPETUAL DAY-DATE II

Reference: 218239

Movement: Automatic, Rolex caliber 3156 (Basis Rolex caliber 3135); Ø 30.97 mm, height 6.47 mm; 31 jewels; 28,800 A/h; Parachrom hairspring, Paraflex shock absorbers, Glucydur balance with Microstella regulating screws; certified chronometer (COSC)

Functions: Hours, minutes, center seconds; date and weekday

Case: White gold, Ø 41 mm, height 11.9 mm; sapphire glass; screw-down crown; water-resistant to 10 bar

Bracelet: President, white gold, folding clasp

Price: $40,700 (€ 30,300)

OYSTER PERPETUAL DAY-DATE II

Reference: 218238

Movement: Automatic, Rolex caliber 3156 (Basis Rolex caliber 3135); Ø 30.97 mm, height 6.47 mm; 31 jewels; 28,800 A/h; Parachrom hairspring, Paraflex shock absorbers, Glucydur balance with Microstella regulating screws; certified chronometer (COSC)

Functions: Hours, minutes, center seconds; date and weekday

Case: Yellow gold, Ø 41 mm, height 11.9 mm; sapphire glass; screw-down crown; water-resistant to 10 bar

Bracelet: President, yellow gold, folding clasp

Price: $38,000 (€ 28,200)

ROLEX MILGAUSS

Magnetic forces are the natural enemy of every mechanical movement. They are found everywhere and can totally disrupt the time display on your wrist. In 1956, watchmakers and engineers from Rolex took a closer look at this problem. They wanted to develop a watch whose inner workings would be completely shielded from external magnetic influences. The result was the first Milgauss, whose main feature is found in its name: It can withstand magnetic disturbances up to 1,000 Gauss and maintain its chronometer accuracy.

The Milgauss has long since become a Rolex classic and over the past 60 years it has continued to be perfected. It is nothing more and nothing less than a modern, high-performance watch with a mechanical heart that can also be relied upon in extreme situations in science and research, like at the CERN nuclear research institute in Geneva, for example.

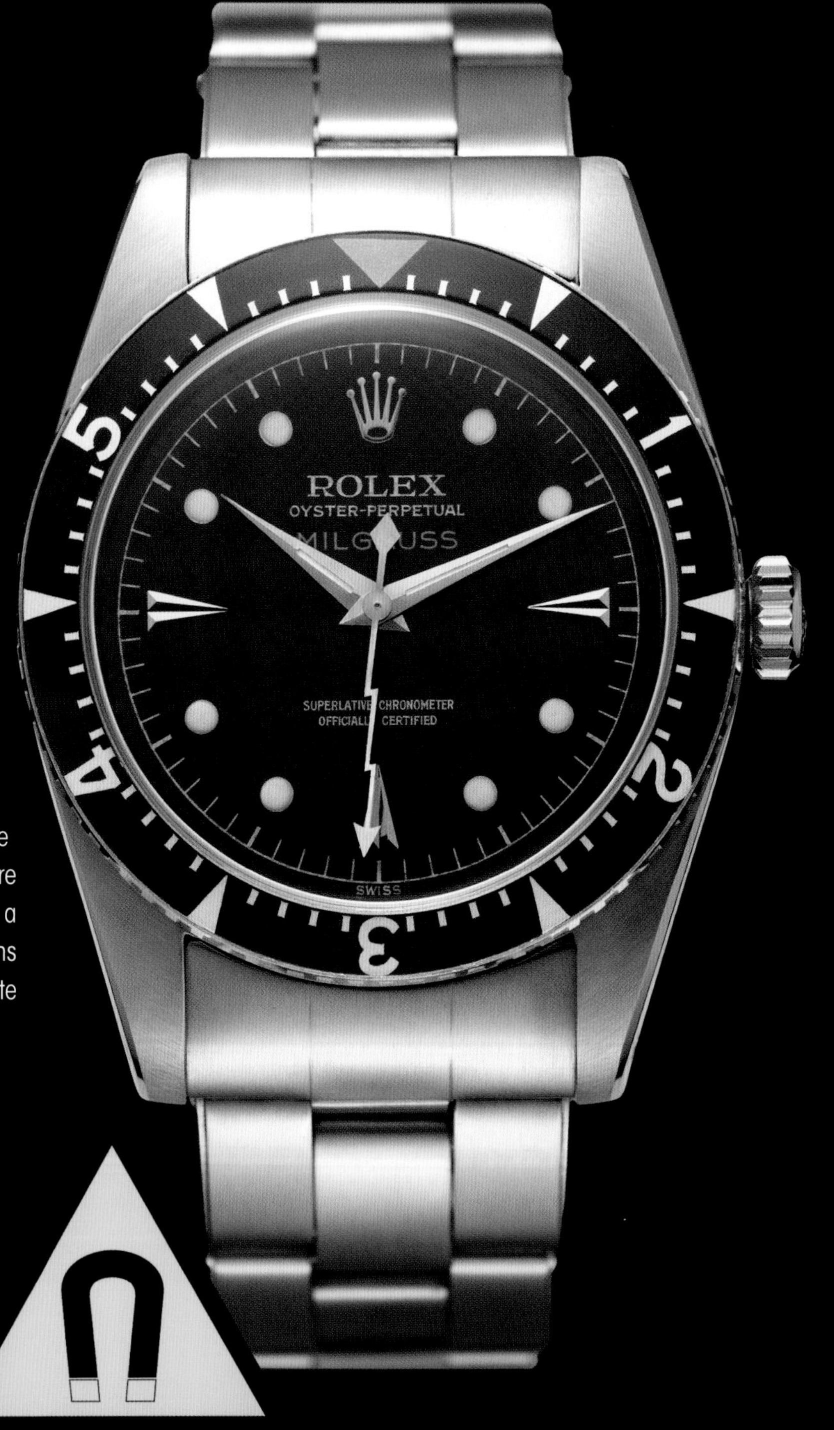

OYSTER PERPETUAL MILGAUSS

Reference:	116400GV
Movement:	Automatic, Rolex caliber 3131 (Basis Rolex caliber 3135); Ø 28.5 mm, height 5.37 mm; 31 jewels; 28,800 A/h; Parachrom hairspring, Paraflex shock absorbers; certified chronometer (COSC); soft iron inner case and dial shield against magnetic fields
Functions:	Hours, minutes, center seconds
Case:	Stainless steel, Ø 40 mm, height 13.2 mm; sapphire glass; screw-down crown; water-resistant to 10 bar
Remarks:	Green tinted sapphire glass
Bracelet:	Oyster, stainless steel, folding clasp with extension link
Price:	$8000 (€ 6000)

OYSTER PERPETUAL MILGAUSS

Reference:	116400
Movement:	Automatic, Rolex caliber 3131 (Basis Rolex caliber 3135); Ø 28.5 mm, height 5.37 mm; 31 jewels; 28,800 A/h; Parachrom hairspring, Paraflex shock absorbers; certified chronometer (COSC); soft iron inner case and dial shield against magnetic fields
Functions:	Hours, minutes, center seconds
Case:	Stainless steel, Ø 40 mm, height 13.2 mm; sapphire glass; screw-down crown; water-resistant to 10 bar
Bracelet:	Oyster, stainless steel, folding clasp with extension link
Price:	$7500 (€ 5590)

ROLEX PRINCE

Within the Rolex model family, the Prince is both its youngest and oldest member. Based on the chronometer from 1928, which was also called the Prince, the re-issued 2005 model with rectangular case combines traditional watchmaking with modern design. The caliber 7040 movement must be wound the traditional way by hand, a deliberate contradiction to the technical automatic models of the Oyster collection. The cases are found exclusively in gold and its various alloys—you also won't find any Prince models in stainless steel—with fine finishing touches on the dial and movement, like the Guilloche patterning for example.

PRINCE

Reference:	5440/8
Movement:	Hand-wound, Rolex caliber 7040-2; 21 jewels; 28,800 A/h; ca. 72-hour power reserve; Glucydur balance with Microstella regulating screws; certified chronometer (COSC)
Functions:	Hours, minutes, auxiliary seconds
Case:	Yellow gold, 45 x 29 mm, height 10 mm; sapphire glass; transparent case back
Bracelet:	Reptile leather, double folding clasp
Price:	$15,800 (€ 11,790)

PRINCE

Reference: 5441/9
Movement: Hand-wound, Rolex caliber 7040-2; 21 jewels; 28,800 A/h; ca. 72-hour power reserve; Glucydur balance with Microstella regulating screws; certified chronometer (COSC)
Functions: Hours, minutes, auxiliary seconds
Case: White gold, 45 x 29 mm, height 10 mm; sapphire glass; transparent case back
Bracelet: Reptile leather, double folding clasp
Price: $15,800 (€ 11,790)

PRINCE

Reference: 5443/8
Movement: Hand-wound, Rolex caliber 7040-2; 21 jewels; 28,800 A/h; ca. 72-hour power reserve; Glucydur balance with Microstella regulating screws; certified chronometer (COSC)
Functions: Hours, minutes, auxiliary seconds
Case: White gold, 45 x 29 mm, height 10 mm; sapphire glass; transparent case back
Bracelet: Reptile leather, double folding clasp
Price: $15,800 (€ 11,790)

TECHNICAL DATA

PRINCE

Reference: 5442/5
Movement: Hand-wound, Rolex caliber 7040-2; 21 jewels; 28,800 A/h; ca. 72-hour power reserve; Glucydur balance with Microstella regulating screws; certified Chronometer (COSC)
Functions: Hours, minutes, auxiliary seconds
Case: Rose gold, 45 x 29 mm, height 10 mm; sapphire glass; transparent case back
Bracelet: Reptile leather, double folding clasp
Price: $15,800 (€ 11,790)

ROLEX

ROLEX FOR LADIES

Women can also appreciate the advantages of timepieces from Geneva and Biel. After all, the mechanical perfection and robustness of the proven Oyster construction merges nicely with feminine elegance and luxury. A mundane wristwatch becomes a real piece of jewelry on the female wrist, and Rolex offers a large selection of model variations to accommodate every taste—no matter how outrageous. Appliqués and cases in fine gold (white or rose gold), brilliantly-cut diamonds or richly colored gemstones set on dials are characteristic of the models in the Lady-Datejust collection, and they're available in different sizes too. The Pearlmaster models, with their opulent gemstones and meticulous mother of pearl dials, represent the crown, so to speak, of women's Rolex models—aside from the fact that they are worn on the wrist and not on the head, their brilliance approaches that of a royal headdress.

OYSTER PERPETUAL LADY-DATEJUST

Reference:	178341
Movement:	Automatic, Rolex caliber 2235 (Basis caliber 2230); Ø 20 mm, height 5.95 mm; 31 jewels; 28,800 A/h; certified chronometer (COSC)
Functions:	Hours, minutes, center seconds; date
Case:	Stainless steel, Ø 31 mm, height 10.5 mm; bezel in rose gold set with 24 diamonds; sapphire glass; screw-down crown; water-resistant to 10 bar
Bracelet:	Oyster, stainless steel and rose gold, folding clasp with extension link
Price:	$15,500 (€ 11,550)

OYSTER PERPETUAL LADY-DATEJUST

Reference:	178240
Movement:	Automatic, Rolex caliber 2235 (Basis caliber 2230); Ø 20 mm, height 5.95 mm; 31 jewels; 28,800 A/h; certified chronometer (COSC)
Functions:	Hours, minutes, center seconds; date
Case:	Stainless steel, Ø 31 mm, height 10.5 mm; sapphire glass; screw-down crown; water-resistant to 10 bar
Bracelet:	Jubilee, stainless steel, folding clasp with extension link
Price:	$6700 (€ 5000)

OYSTER PERPETUAL LADY-DATEJUST

Reference:	179383
Movement:	Automatic, Rolex caliber 2235 (Basis caliber 2230); Ø 20 mm, height 5.95 mm; 31 jewels; 28,800 A/h; certified chronometer (COSC)
Functions:	Hours, minutes, center seconds; date
Case:	Stainless steel, Ø 26 mm, height 10 mm; bezel in yellow gold set with 46 diamonds
Remarks:	Dial set with 10 diamonds; water-resistant to 10 bar
Bracelet:	Jubilee, stainless steel and yellow gold, folding clasp
Price:	$20300 (€ 15,150)

OYSTER PERPETUAL LADY-DATEJUST

Reference: 178288
Movement: Automatic, Rolex caliber 2235 (Basis caliber 2230); Ø 20 mm, height 5.95 mm; 31 jewels; 28,800 A/h; certified chronometer (COSC)
Functions: Hours, minutes, center seconds; date
Case: Yellow gold, Ø 31 mm, height 10.5 mm; bezel set with 48 diamonds; sapphire glass; screw-down crown; water-resistant to 10 bar
Remarks: Dial set with 8 diamonds, numbers "6" and "9" inlaid with rubies
Bracelet: Oyster, yellow gold, folding clasp with extension link
Price: $43,400 (€ 32,350)

OYSTER PERPETUAL LADY-DATEJUST

Reference: 178278
Movement: Automatic, Rolex caliber 2235 (Basis caliber 2230); Ø 20 mm, height 5.95 mm; 31 jewels; 28,800 A/h; certified chronometer (COSC)
Functions: Hours, minutes, center seconds; date
Case: Yellow gold, Ø 31 mm, height 10.5 mm; sapphire glass; screw-down crown; water-resistant to 10 bar
Remarks: Dial set with 8 diamonds, number "6" inlaid with rubies
Bracelet: President, yellow gold, folding clasp
Price: $30,800 (€ 22,950)

TECHNICAL DATA

OYSTER PERPETUAL DATEJUST SPECIAL EDITION

Reference: 81338
Movement: Automatic, Rolex caliber 2235 (Basis caliber 2230); Ø 20 mm, height 5.95 mm; 31 jewels; 28,800 A/h; certified chronometer (COSC)
Functions: Hours, minutes, center seconds; date
Case: Yellow gold, Ø 34 mm, height 10.5 mm; bezel set with 116 diamonds; sapphire glass; screw-down crown; water-resistant to 10 bar
Remarks: Dial with diamond pavés
Bracelet: Pearlmaster, yellow gold, set with 262 diamonds, folding clasp
Price: $98,300 (€ 73,200)

OYSTER PERPETUAL LADY-DATEJUST PEARLMASTER

Reference: 80299
Movement: Automatic, Rolex caliber 2235 (Basis caliber 2230); Ø 20 mm, height 5.95 mm; 31 jewels; 28,800 A/h; certified chronometer (COSC)
Functions: Hours, minutes, center seconds; date
Case: White gold, Ø 29 mm, height 10 mm; bezel set with 32 diamonds; sapphire glass; screw-down crown; water-resistant to 10 bar
Remarks: Mother of pearl dial set with 10 diamonds
Bracelet: Pearlmaster, white gold, set with 174 diamond, folding clasp
Price: $74,500 (€ 55,550)

OYSTER PERPETUAL LADY-DATEJUST PEARLMASTER

Reference:	80285
Movement:	Automatic, Rolex caliber 2235 (Basis caliber 2230); Ø 20 mm, height 5.95 mm; 31 jewels; 28,800 A/h; certified chronometer (COSC)
Functions:	Hours, minutes, center seconds; date
Case:	Rose gold, Ø 29 mm, height 10 mm; bezel set with 32 diamonds; sapphire glass; screw-down crown; water-resistant to 10 bar
Remarks:	Mother of pearl dial, numeral "6" inlaid with 11 diamonds
Bracelet:	Pearlmaster, rose gold and white gold, set with 226 diamonds, folding clasp
Price:	Upon request

OTHER BOOKS IN THIS SERIES

Breitling Highlights
by Henning Mützlitz.
ISBN 978-0-7643-4211-0. $29.99

A. Lange & Söhne® Highlights
by Henning Mützlitz.
ISBN 978-0-7643-4361-0. $29.99

Omega Highlights
by Henning Mützlitz.
ISBN 978-0-7643-4212-7. $29.99

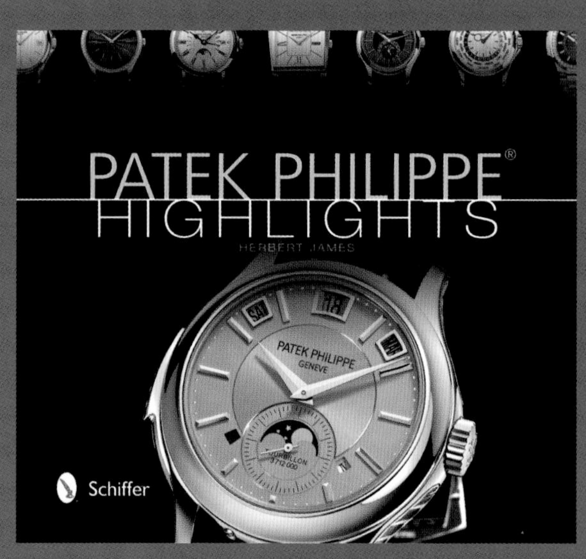

Patek Philippe® Highlights
by James Herbert
ISBN 978-0-7643-4322-3. $29.99